Red Hat Certified Specialist in Services Management and Automation EX358 Exam Guide

Get your certification and prepare for real-world challenges
as a Red Hat Certified Specialist

Eric McLeroy

BIRMINGHAM—MUMBAI

Red Hat Certified Specialist in Services Management and Automation EX358 Exam Guide

Group Product Manager: Mohd Riyan Khan

Publishing Product Manager: Prachi Sawant

Content Development Editor: Sujata Tripathi

Technical Editor: Shruthi Shetty

Copy Editor: Safis Editing

Project Coordinator: Ashwin Dinesh Kharwa

Proofreader: Safis Editing

Indexer: Rekha Nair

Production Designer: Vijay Kamble

Senior Marketing Coordinator: Marylou De Mello

First published: February 2023

Production reference: 1020223

Published by Packt Publishing Ltd.

Livery Place

35 Livery Street

Birmingham

B3 2PB, UK.

ISBN 978-1-80323-549-3

www.packtpub.com

To my wife, Jennifer – you are always by my side when I choose to take on what appear to be impossible tasks. With your and our children's help, I always seem to manage to do the impossible. Thank you.

– Eric McLeroy

Foreword

Ever since its inception, Ansible has been billed as *simple, powerful, agentless* automation software. One of the defining characteristics of Ansible has always been its emphasis on delivering practical, automated solutions to everyday operational challenges. In order to accomplish this goal, it has always been paramount to the foundation of Ansible to adapt and deliver solutions that advance an organization's ability to quickly and efficiently manage its infrastructure while staying focused on solving everyday, real-world problems.

I have benefitted from the opportunity to work closely with Eric McLeroy from the time he was hired by Red Hat back in 2016. His work has continuously focused on building Ansible solutions, helping customers realize the full benefit of automation in their own infrastructures through measurable results.

During this time, Eric has brought a key perspective, based on his own experiences, as to the challenges facing IT organizations looking to embrace automation and solve the operational challenges associated with today's complex heterogeneous environments. He has worked tirelessly to share his knowledge and expertise, continuing to find innovative ways to apply Ansible technology while staying true to Ansible's philosophy around simplicity.

Over the course of this book, Eric has delivered a masterful set of examples that remains true to the origins and defining characteristics that made Ansible one of the industry-leading automation platforms available today. By relying on this extensive background in automating critical IT infrastructure, Eric has captured, throughout the course of this book, a diverse set of real-life challenges with practical, proven techniques using Ansible to build and deploy automated, scalable solutions.

With his guidance and approach captured in the chapters of this book, Eric has woven together comprehensive scenarios including screenshots, step-by-step examples, and clear instructions that will be as beneficial to users just starting their Ansible journey as for users with years of experience with Ansible.

Start automating today with Ansible.

Peter Sprygada

Former Distinguished Engineer and Chief Architect, Ansible Automation Platform

Contributors

About the author

Eric McLeroy brings over 15 years of technical expertise and experience in virtualization, system administration, networking, and automation to his educational works. In addition to his field experience and certifications, Eric has three master's degrees.

Eric knows what it is like to be a certification student, as he holds 18 certifications from Red Hat, Cisco, VMware, and Microsoft. This experience allows him to understand the experience from a student's perspective and focus his educational materials on the needs of those aiming to be certified. Eric also brings his first-hand experience from the tech trenches to ensure his works are practical and will help in your everyday activities of working on mission-critical production systems.

About the reviewers

Tomoya Ogawa has been working as a hybrid IT infra and security engineer for a global pharmaceutical company's Japan subsidiary since 2022 in order to launch new digital solutions. Previously, he worked as an IT architect of Linux, security, network, and Microsoft solutions for IBM/Kyndryl-Japan and NTT-West for over 15 years.

First and foremost, I would like to thank Prathamesh Walse and Ashwin Dinesh Kharwa of Packt Publishing for assigning a relatively lesser known figure in the industry as a technical reviewer and taking a chance on me. Secondly, I would like to thank my family for their support of my weekend challenges.

Krystian Majda is a Linux expert with more than 10 years of experience and several certifications. He holds an active Red Hat Certified Architect title (certification ID: 140-182-569). Krystian has worked as a subject matter expert for years and seeks to share his knowledge with others. With a strong DevOps approach to problems and tasks, he automates as much as possible and likes to collaborate in infrastructure-as-code ecosystems – always with security in mind. Krystian is a freelancer and Red Hat Certified Specialist in Services Management and Automation (the EX358 exam), which is the main subject of this book. He is not afraid of new challenges and is constantly building his skillset.

Table of Contents

Preface   xiii

Part 1: Red Hat Linux 8 –Configuring and Maintaining Storage with Automation

1

Block Storage – Learning How to Provision Block Storage on Red Hat Enterprise Linux

Block Storage – Learning How to Provision Block Storage on
Red Hat Enterprise Linux   3

Technical requirements	4	iSCSI block storage – manual
Setting up GitHub access	4	provisioning and deployment 19
Setting up your lab environment	4	iSCSI block storage – Ansible
iSCSI block storage – overview of		automation playbook creation
what it is and why we need it	16	and usage 27
Testing your knowledge	17	Summary 31

2

Network File Storage – Expanding Your Knowledge of How to Share Data

Network File Storage – Expanding Your Knowledge of
How to Share Data   33

Technical requirements	34	NFS Ansible Automation
Setting up GitHub access	34	playbook creation and usage 41
Setting up your lab environment		SMB storage manual provisioning
for NFS and SMB	34	and deployment 43
NFS and SMB network storage –		SMB storage Ansible Automation
the way they work and when to		playbook creation and usage 49
choose one over the other	35	Summary 54

Part 2: Red Hat Linux 8 – Configuring and Maintaining Networking with Automation

3

Network Services with Automation – Introduction to Red Hat Linux Networking   57

Technical requirements	57	Getting to know what the different terms mean and how they apply to what you are trying to achieve!	58
Setting up GitHub access	58		
Setting up your lab environment for networking	58	Creating your basic network profile	59
The beginning of your journey into Linux networking!	58	Automation of network services using Ansible	70
		Summary	78

4

Link Aggregation Creation – Creating Your Own Link and Mastering the Networking Domain   79

Technical requirements	80	Creating different types of link aggregation profiles	84
Setting up GitHub access	80		
Setting up your lab environment for Network Interface Card (NIC) teaming	80	Taking the headache out of setting up your link aggregation by using Ansible Automation	97
Getting to know link aggregation	83	Summary	104

5

DNS, DHCP, and IP Addressing – Gaining Deeper Knowledge of Red Hat Linux Networking   105

Setting up GitHub access	106	Diving deeper into Linux networking where we look at DNS, DHCP, and static IP addressing	106

Setting up static IP addresses
for times when DHCP is
not available but you still need
to make that service reachable   106

Using the basic out-of-the-box
DHCP configuration to get online
fast when available on your network   113

Using DHCP at initial interface connectivity
provided by an external source   113

Setting up DHCP server configuration
manually to provide DHCP services   119

Automating DHCP server configuration
to provide DHCP services   128

Learning about the DNS and
why you need to know about it   133

Setting up DNS server configuration
manually to provide DNS services   133

Automating DNS server configuration
to provide DNS services   147

Summary   154

Part 3: Red Hat Linux 8 – Configuring and Maintaining Applications with Automation and a Comprehensive Review with Exam Tips

6

Printer and Email – Setting Up Printers and Email Services on Linux Servers   157

Technical requirements   158

Setting up GitHub access   158

Learning about printer services
and setting them up manually   158

Setting up printer services via
Ansible Automation   169

Learning about email services
and setting them up manually   176

Setting up email services via
Ansible Automation   182

Summary   186

7

Databases – Setting Up and Working with MariaDB SQL Databases   187

Technical requirements   187

Setting up GitHub access   188

Getting started with MariaDB for
data collection storage   188

Installing and configuring
MariaDB on RHEL 8 manually   188

Installing and configuring
MariaDB on RHEL 8 using
Ansible Automation   209
Summary   217

8

Web Servers and Web Traffic – Learning How to Create and Control Traffic   219

Technical requirements   219
Setting up GitHub access   220

Getting started with web servers
and traffic control   220

Learning to set up web servers
manually and control traffic   220

Learning how to use Ansible
Automation to automate
web servers and control traffic   240
Summary   248

9

Comprehensive Review and Test Exam Questions   249

Technical requirements   249
Setting up GitHub access   249

A comprehensive review of
all exam objectives and
mock exams for you to test
your newfound skills   250
Managing network services   251
Managing firewall services   260
Managing SELinux   265

Managing system processes   267
Managing link aggregation   270
Managing DNS   274
Managing DHCP   283
Managing printers   286
Managing email services   289
Managing a MariaDB database server   291
Managing web access   295
Managing NFS   298

Summary   305

10

Tips and Tricks to Help with the Exam   307

Technical requirements   307
Scheduling your exam   307

Tips and tricks that can
help while taking the test   315
Summary   321

Index 323

Other Books You May Enjoy 328

Preface

This book explores the world of Red Hat Linux Server Management and Automation in an effort to help you secure success when taking the EX358 certification exam. It goes into detail about each of the main objectives laid out by Red Hat in order to fully prepare you to complete this step in your career journey.

Applying many years of experience with Red Hat and Linux in general, I will delve into the different applications that you need to know about in order to help you understand Red Hat Enterprise Linux and Ansible Automation. I have worked for Red Hat for over 6 years and on Red Hat systems for well over 10. I hold many certifications, including Red Hat Certified Architect, obtaining which encompasses multiple Red Hat products and disciplines.

This book will help you face all the nuances that you will run into in the testing environment and cover helpful areas that can save you time and effort, thus guiding you to complete this certification exam with top marks. Each objective is broken down and laid out in a meaningful way that will help you retain the knowledge needed to excel in this exam.

Who this book is for

This book is for Linux system administrators working to advance their careers. This book will help you gain additional skills that will enable you to pass this industry exam. You should have a strong background in Red Hat Linux and Ansible Automation before undertaking the tasks laid out in this book.

What this book covers

Chapter 1, Block Storage – Learning How to Provision Block Storage on Red Hat Enterprise Linux, goes into detail on the block storage configuration of an iSCSI setup and its connectivity.

Chapter 2, Network File Storage – Expanding Your Knowledge of How to Share Data, explains the setup of NFS and Samba, along with the permissions and connectivity to shared drives.

Chapter 3, Network Services with Automation – Introduction to Red Hat Linux Networking, marks the beginning of working with Red Hat Enterprise Linux networking, such as IP addressing.

Chapter 4, Link Aggregation Creation – Creating Your Own Link and Mastering the Networking Domain, explains how to set up a network team on a Red Hat Enterprise Linux server.

Chapter 5, DNS, DHCP, and IP Addressing – Gaining Deeper Knowledge of Red Hat Linux Networking, delves into setting up and providing DNS and DHCP services to your infrastructure from a Red Hat Enterprise Linux server.

Chapter 6, Printer and Email – Setting Up Printers and Email Services on Linux Servers, covers how to set up printing services, as well as a null email client.

Chapter 7, Databases – Setting Up and Working with MariaDB SQL Databases, teaches you how to install, secure, and configure MariaDB using MySQL commands on Red Hat Enterprise Linux servers.

Chapter 8, Web Servers and Web Traffic – Learning How to Create and Control Traffic, goes over how to install and configure Apache and Nginx web servers. We will also go into how to control traffic to the web servers connected to our infrastructure.

Chapter 9, Comprehensive Review and Test Exam Questions, presents some example questions so that we can apply everything that we have learned throughout the book.

Chapter 10, Tips and Tricks to Help with the Exam, looks at some simple tips and tricks that can save you time and increase your success rate when taking the exam.

To get the most out of this book

You should have a working knowledge of Red Hat Enterprise Linux and you should have written playbooks in Ansible Automation before when you pick up this book. Without these skills, this book and exam might be daunting to you.

Software/hardware covered in the book	Operating system requirements
Ansible 2.9	Red Hat Enterprise Linux 8.1
VirtualBox	OSX or Windows

You will need internet connectivity from your workspace in order to download the needed software. You will need to utilize VirtualBox to follow along with the lab layout and understand that there may be differences if you deviate from it. You will need to set up a developer license for Red Hat, instructions for which are laid out in *Chapter 1*.

If you are using the digital version of this book, we advise you to type the code yourself or access the code from the book's GitHub repository (a link is available in the next section). Doing so will help you avoid any potential errors related to the copying and pasting of code.

You will need to have experience running Red Hat Enterprise Linux, as well as familiarity with Ansible Automation. It is recommended to have real-life experience of these systems and applications in order to make the exam easier to follow and understand.

Download the example code files

You can download the example code files for this book from GitHub at `https://github.com/PacktPublishing/Red-Hat-Certified-Specialist-in-Services-Management-and-Automation-EX358-Exam-Guide`. If there's an update to the code, it will be updated in the GitHub repository.

We also have other code bundles from our rich catalog of books and videos available at `https://github.com/PacktPublishing/`. Check them out!

Download the color images

We also provide a PDF file that has color images of the screenshots and diagrams used in this book. You can download it here: `https://packt.link/Yrj1x`.

Conventions used

There are a number of text conventions used throughout this book.

`Code in text`: Indicates code words in text, database table names, folder names, filenames, file extensions, pathnames, dummy URLs, user input, and Twitter handles. Here is an example: "Keep in mind that your `/etc/hosts` file will look different based on your IPs."

A block of code is set as follows:

```
tasks:
    - name: Install Samba and Samba-Client
      package:
        name:
            - samba
            - samba-client
            - cifs-utils
        state: latest
```

Any command-line input or output is written as follows:

```
$ ssh-keygen
```

Bold: Indicates a new term, an important word, or words that you see onscreen. For instance, words in menus or dialog boxes appear in **bold**. Here is an example: "Choose **Edit a connection**."

> **Tips or important notes**
> Appear like this.

Get in touch

Feedback from our readers is always welcome.

General feedback: If you have questions about any aspect of this book, email us at customercare@packtpub.com and mention the book title in the subject of your message.

Errata: Although we have taken every care to ensure the accuracy of our content, mistakes do happen. If you have found a mistake in this book, we would be grateful if you would report this to us. Please visit www.packtpub.com/support/errata and fill in the form.

Piracy: If you come across any illegal copies of our works in any form on the internet, we would be grateful if you would provide us with the location address or website name. Please contact us at copyright@packtpub.com with a link to the material.

If you are interested in becoming an author: If there is a topic that you have expertise in and you are interested in either writing or contributing to a book, please visit authors.packtpub.com.

Share Your Thoughts

Once you've read *Red Hat Certified Specialist in Services Management and Automation EX358 Exam Guide*, we'd love to hear your thoughts! Scan the QR code below to go straight to the Amazon review page for this book and share your feedback.

https://packt.link/r/1803235497

Your review is important to us and the tech community and will help us make sure we're delivering excellent quality content.

Download a free PDF copy of this book

Thanks for purchasing this book!

Do you like to read on the go but are unable to carry your print books everywhere?

Is your eBook purchase not compatible with the device of your choice?

Don't worry, now with every Packt book you get a DRM-free PDF version of that book at no cost.

Read anywhere, any place, on any device. Search, copy, and paste code from your favorite technical books directly into your application.

The perks don't stop there, you can get exclusive access to discounts, newsletters, and great free content in your inbox daily

Follow these simple steps to get the benefits:

1. Scan the QR code or visit the link below

https://packt.link/free-ebook/9781803235493

2. Submit your proof of purchase
3. That's it! We'll send your free PDF and other benefits to your email directly

Part 1: Red Hat Linux 8 –Configuring and Maintaining Storage with Automation

In this part, you will learn to set up and maintain block and file storage manually and automatically. This will meet the objectives set to pass the Red Hat EX358 exam.

This part contains the following chapters:

- *Chapter 1, Block Storage – Learning How to Provision Block Storage on Red Hat Enterprise Linux*
- *Chapter 2, Network File Storage – Expanding Your Knowledge of How to Share Data*

1

Block Storage – Learning How to Provision Block Storage on Red Hat Enterprise Linux

Block storage within **Red Hat Enterprise Linux** (**RHEL**) makes up the foundation of many core applications. You will use it for many things within the world of Linux, from application development to backups to deployments of infrastructure such as OpenStack using **Internet Small Computer Systems Interface** (**iSCSI**). Through understanding how and when to use block storage over other storage options and how to provision it through manual steps as well as automate it through Ansible, you will be able to comprehend and grasp the knowledge needed for your day-to-day work with Linux as well as ensuring you understand the building blocks required to meet the needs of the *EX358* exam. These lessons not only allow you to complete the *EX358* exam with success but also enable you to better understand why we use block storage over other filesystems in situations that dictate the use of this filesystem type in real-world scenarios.

This comes in handy when you are building your infrastructure at your company, in your home lab for learning purposes, or for that start-up you always wanted to create. At the end of this chapter, you will be able to provision block storage using Red Hat best practices both manually and through Ansible automation in order to meet the requirements of Red Hat. This will allow you to gain support from Red Hat if you have an active contract and also gain help from the community if you do not have Red Hat support in order to resolve any issues you may run into during your usage of this technology.

You will be able to configure iSCSI initiators, boot with them both manually and through Ansible automation, and safely tear down unused variations of the iSCSI block storage after you are done with this chapter. This will, in turn, ensure your full understanding of the overall life cycle and the effective nature of block storage in your ecosystem.

In this chapter, we're going to cover the following main topics:

- iSCSI block storage—overview of what it is and why we need it
- iSCSI block storage—manual provisioning and deployment
- iSCSI block storage—Ansible automation playbook creation and usage

Technical requirements

Before we delve into the topics in detail, you will need to set a few things up. Let's look at what they are.

Setting up GitHub access

You will need a free GitHub account in order to access some of the code that will be provided throughout this book. Please sign up for a free account at https://github.com/. We will be utilizing the code found in the following repository throughout the course of this book: https://github.com/PacktPublishing/Red-Hat-Certified-Specialist-in-Services-Management-and-Automation-EX358-. We will be utilizing the code snippets found in the ch1 folder of this code repository (aka repo) for our iSCSI automation hands-on exercises, which can be found here: https://github.com/PacktPublishing/Red-Hat-Certified-Specialist-in-Services-Management-and-Automation-EX358-Exam-Guide/tree/main/Chapter01. The code placed here will allow you to check your work and ensure you are on the right track when writing your playbooks within Ansible. Please keep in mind these are one person's way of writing tested playbooks that will meet the exam objectives; however, there are many ways of writing successful playbooks to meet these objectives.

Setting up your lab environment

All of the demonstrations of VirtualBox and coding will be shown on macOS but can be performed on Windows as well as Linux OSs. We will be setting up some iSCSI block devices. First, you will need a machine that can run VirtualBox with enough memory to run your machine and three VMs that each have 2 GB of memory, one 10 GB hard drive, and one 5 GB hard drive, which equals 15 GB of required hard drive space per VM, as can be seen in the following screenshot:

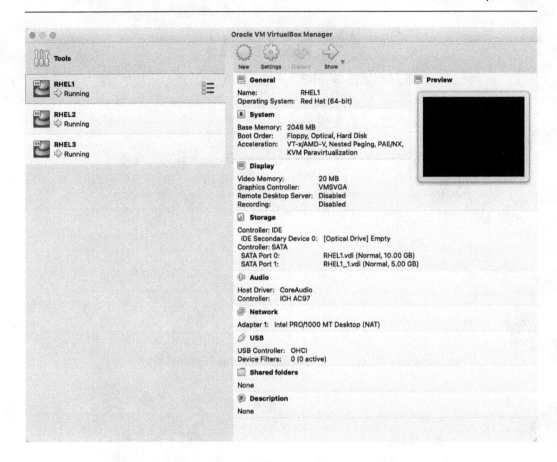

Figure 1.1 – Layout of the VirtualBox deployment

This is mainly for the storage hands-on labs, and you can revert to one 10 GB hard drive for exercises. RHEL 8.1 requires at least 9.37 GB of space to run. Using a Red Hat Developer account (https://developers.redhat.com/), you can access real Red Hat software to develop your skills as well as the software in order to set this up:

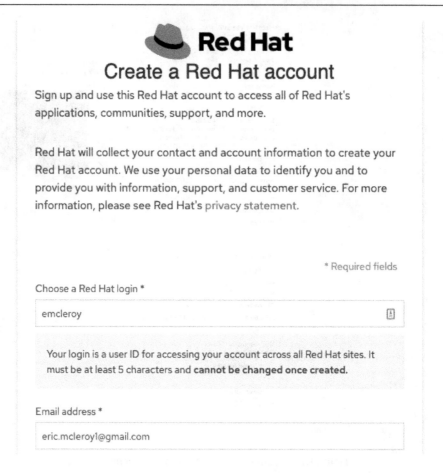

Figure 1.2 – Signup is simple!

Because the exam is set for RHEL 8.1, I recommend using this version for your studying needs in order to get the most authentic exam-like infrastructure possible. In the following screenshot, the correct version you should download is the first option:

8.1.0

RHEL8 x86_64	DVD iso	Release date November 05, 2019	Download (7.31 GB)
RHEL8 x86_64	Boot iso	Release date November 05, 2019	Download (564 MB)
RHEL8 aarch64	DVD iso	Release date November 05, 2019	Download (5.14 GB)
RHEL8 aarch64	Boot iso	Release date November 05, 2019	Download (526.46 MB)

Figure 1.3 – The correct version for the exam and for you

This will be true for the entirety of the book, including the comprehensive review and lab at the end. Before installing the OS, you can create a second hard drive in VirtualBox from the settings, as can be seen in the following screenshot:

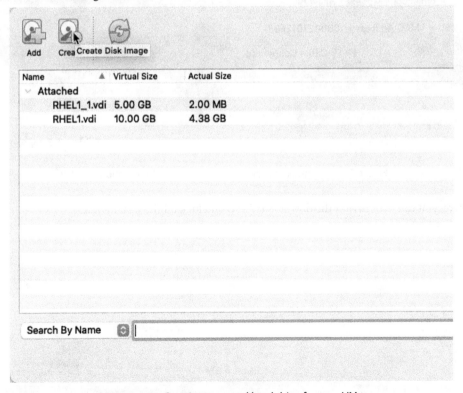

Figure 1.4 – Creating a second hard drive for your VM

You also need to ensure that you choose **Bridged Adapter** mode for your network **Attached to** option. The **Promiscuous Mode** option is also allowed so that it can reach the internet and other adapters. One caveat to keep in mind is that bridged-over Wi-Fi does not always play nice, so try to ensure you have a wired connection if you are setting up your lab in this manner:

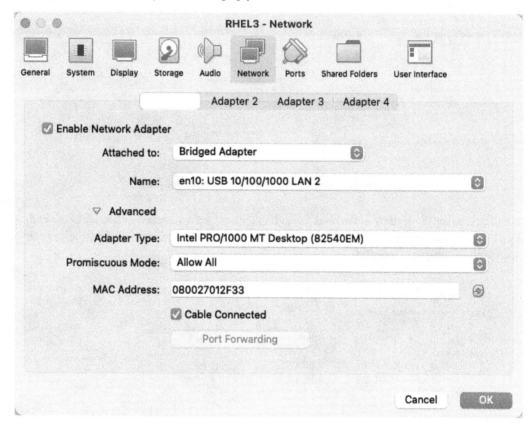

Figure 1.5 – Bridged adapter with Promiscuous Mode option

From here, you can then mount the downloaded ISO and kick off the installation:

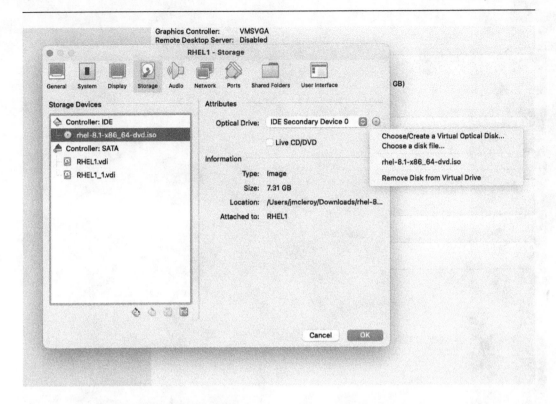

Figure 1.6 – Mounting RHEL DVD ISO that was downloaded previously

There are some best practices you need to keep in mind. We will be installing the **Server with the GUI** option. Make sure to create yourself an administrator account as well as keeping your root account as you will want to do everything as sudo and not directly as root for security purposes and all-around good habits. The user creation screen, as follows, allows you to set up your root password and any users you would like to create:

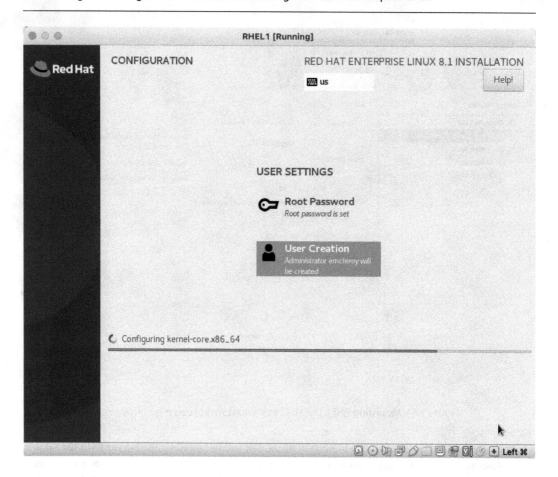

Figure 1.7 – Administrator accounts are best practices; sudo over root is always preferred

Next, you will need to use the login for your Red Hat Developer account and license the VMs using the account credentials. See the following screenshot for how to correctly apply a Red Hat subscription license:

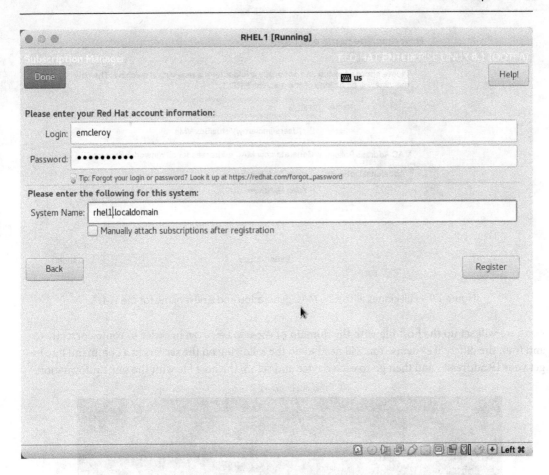

Figure 1.8 – Red Hat Developer credentials or an active Red Hat account needed

You can create one machine and then clone it into the other two you need. Make sure you choose to generate new MAC addresses and to make a full clone to ensure that no overlap causes network or storage issues, as shown in the following screenshot:

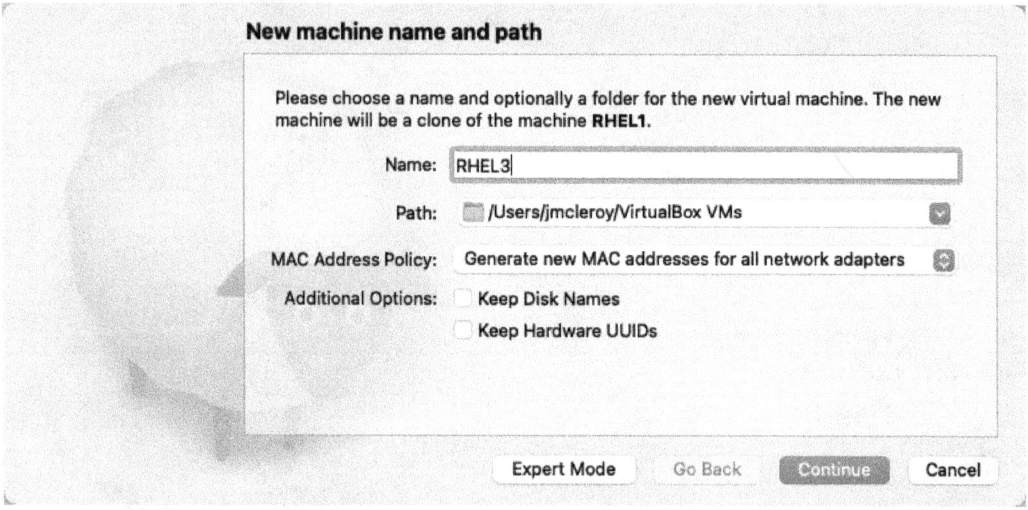

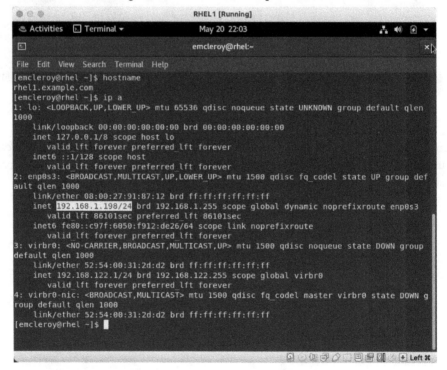

Figure 1.9 – Full clones with new MAC generation and a new name for the VM

Next, we will set up the host file with the domain of example.com in order to route correctly to and from the different systems. You will need to do the following on the servers in a command line to get your IP addresses and then go to each device and set up the host file with the same information:

Figure 1.10 – Hostname and IP of rhel1.example.com system

Next, let's gather the hostnames or change them to what you would like them to be using the following commands and review the output in this case, which is `rhel1.example.com`:

```
[emcleroy@rhel1 ~]$ sudo hostnamectl set-hostname rhel1.
example.com
[emcleroy@rhel1 ~]$ hostname
 rhel1.example.com
```

Use the hostnames and the IP addresses to build the inventory for the host file. After you do this, make sure that you shut down the system for it to save the changes permanently. Next, you're going to want to add these as noted to the host file on all three VMs using the following command:

```
$ sudo vi /etc/hosts
```

Here is an example of the completed `/etc/hosts` file:

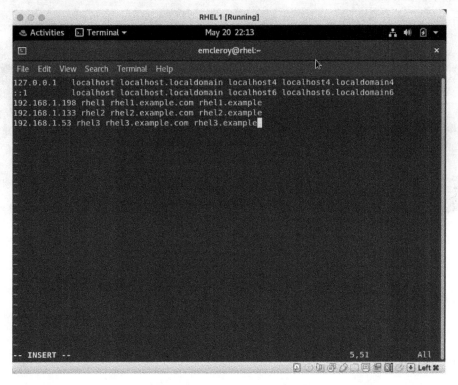

Figure 1.11 – Finished /etc/hosts file

Keep in mind your /etc/hosts file will look different based on your IPs. You should now be able to ping via the hostname and IP of all of the different VMs from one to another:

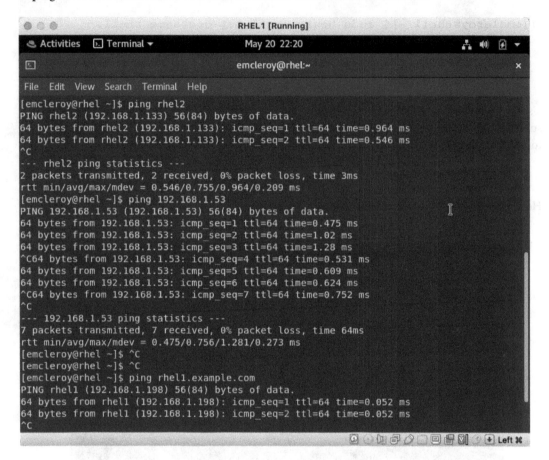

Figure 1.12 – Example of working networking environment

Next, for ease of use, let's set up passwordless sudo for our user account, which in my case would be emcleroy.

We will start by running the following command:

```
$ sudo visudo
```

Next, we will locate the lines of code highlighted in the following screenshot and add the highlighted lines of text. Also, note that if you are allowing administrators, you can simply uncomment # in front of the %wheel line as well:

```
● ● ●          jmcleroy — emcleroy@rhel2:~/Documents — ssh emcleroy@rhel2 — 80×25
##         user     MACHINE=COMMANDS
##
## The COMMANDS section may have other options added to it.
##
## Allow root to run any commands anywhere
root     ALL=(ALL)        ALL
ALL      ALL=(ALL)        ALL
## Allows members of the 'sys' group to run networking, software,
## service management apps and more.
# %sys ALL = NETWORKING, SOFTWARE, SERVICES, STORAGE, DELEGATING, PROCESSES, LOC
ATE, DRIVERS

## Allows people in group wheel to run all commands
%wheel  ALL=(ALL)        ALL

## Same thing without a password
# %wheel          ALL=(ALL)        NOPASSWD: ALL
emcleroy          ALL=(ALL)        NOPASSWD: ALL

## Allows members of the users group to mount and unmount the
## cdrom as root
# %users  ALL=/sbin/mount /mnt/cdrom, /sbin/umount /mnt/cdrom

## Allows members of the users group to shutdown this system
-- INSERT --
```

Figure 1.13 – Highlighted lines of text to be added, substituting your username for mine

You will need to do this for all three of the servers.

Finally, we will add SSH keys across the servers to allow for fast connectivity so that we do not have to type passwords every time we need to log in from one server to another. Start by generating SSH keys with the following command on your `rhel1` VM:

```
$ ssh-keygen
```

Just leave the defaults and keep hitting *Enter*, and then once that is generated, you will want to do the following:

```
$ ssh-copy-id -i ~/.ssh/id_rsa.pub username@server
```

This will push the keys to the servers and allow all the servers to talk bi-directionally. You will want to do that for all three servers, so you will do the following (including for the server you are currently on to ensure that the keys are pushed to the known host file for all of the servers):

```
$ ssh-copy-id -i ~/.ssh/id_rsa.pub emcleroy@rhel1
$ ssh-copy-id -i ~/.ssh/id_rsa.pub emcleroy@rhel2
$ ssh-copy-id -i ~/.ssh/id_rsa.pub emcleroy@rhel3
```

From here, you have full access to a three-VM lab running RHEL 8.1 with secondary HDDs for use with this iSCSI hands-on exercise. The only minor differences will come up in the networking hands-on labs where we will go over adding additional **network interface controllers** (**NICs**) for network teaming. This will be another topic that you need to understand in order to ace the *EX358* exam.

Congratulations! You have now successfully set up your lab environment. Pat yourself on the back and take a break. We will now be talking about the manual steps to build out iSCSI block devices and use them. This will be followed by putting that lab environment you just built to the test and getting hands-on experience with the technology.

iSCSI block storage – overview of what it is and why we need it

There are a number of things you need to know about block storage and, in this case, iSCSI. It is a **storage area network** (**SAN**) protocol that allows for devices or parts of devices to be seen as block storage by an end device. SAN is how iSCSI connects to the network and gives the ability to provide network **logical unit numbers** (**LUNs**). This allows systems to use these block devices as if they were physical hard drives in the system that they can boot from, save files to, or use like any hard drive that you have in your normal computer. With this in mind, we have to take a few things into account.

First, you have to ensure that your network can handle the connectivity without congestion as this will cause your systems to slow down and possibly lag behind what you are doing, causing users to become frustrated. Knowing this, you have to plan out your SAN extremely well and properly network out your block storage onto a normally non-encrypted network setup that meets the minimum speeds of 10 GB but can go much higher in a lot of cases. This allows smooth usage of your storage without the headaches you will run into as a system administrator. If you were to put this on the same network as your LAN traffic and expect your streaming (don't do this while at work!) users are watching videos while trying to also do their jobs from a machine that is hosted from a SAN iSCSI block storage device. Other things to keep in mind are you need to ensure proper `firewalld` syntax is utilized and SELinux protocols are followed to allow connectivity at startup or you will have a giant paperweight without much happening.

There are some main items you have to take into account when you are looking at iSCSI using `targetcli`, and I will get to more details about `targetcli` as that is the toolset we will utilize to allow us to use iSCSI in our RHEL 8.1 environment. The main things you need to know are the

initiator, **target**, **Portal**, **LUN**, **Access Control List** (**ACL**), and **Target Portal Group** (**TPG**). These items make up iSCSI storage and lead to a lot of misconceptions. Let's test your knowledge before we dig deeper into the systems and how they work together to provide block storage over the network to remote servers.

Testing your knowledge

Answer the following questions:

1. What is an iSCSI storage source on an iSCSI server?

    A. Target

    B. LUN

    C. **iSCSI Qualified Name (IQN)**

    D. ACL

2. What is a unique worldwide name used to identify both initiators and targets?

    A. Target

    B. LUN

    C. IQN

    D. ACL

3. An iSCSI client that is typically software-based is known as a:

    A. TPG

    B. Portal

    C. IQN

    D. Initiator

4. Which of the following is an access restriction using the IQN?

    A. Target

    B. LUN

    C. IQN

    D. ACL

5. What is the most commonly used software for setting up RHEL 8.1 iSCSI block storage?

    A. `firewalld`

    B. SELinux

    C. `targetcli`

    D. `networkd`

6. Which service or port do you need to allow for iSCSI to work through your firewall?

    A. `iSCSI-target`

    B. `3260/UDP`

    C. `iSCSI`

    D. `targetcli`

7. What includes the named item `2020-06.com.mcleroy.www`?

    A. Target

    B. LUN

    C. IQN

    D. ACL

8. Which system do you need to enable to ensure iSCSI will start at boot?

    A. `firewalld`

    B. Target

    C. `targetcli`

    D. `networkd`

Answers:

1. A. Target

2. C. IQN

3. D. Initiator

4. D. ACL

5. C. `targetcli`

6. A. `iSCSI-target`

7. C. IQN

8. B. Target

iSCSI block storage – manual provisioning and deployment

We will start by installing `targetcli` and using that to set up iSCSI to provide block-based storage to other systems for file usage, boot systems, and so on. This will showcase the wide range of uses that come with iSCSI block storage implemented with RHEL 8.1. We will then show how to decommission the storage device and clean up the systems after utilizing the resources.

First, we will install `targetcli` in order to utilize the iSCSI systems on `rhel1`:

```
$ sudo dnf install targetcli -y
```

We will follow that by enabling the system to start up the iSCSI block storage. When the system boots or has an issue that causes the target system to need to restart, it will reload the service in order to keep the storage up and operational:

```
$ sudo systemctl enable target
```

After that, we will allow `iscsi-initiator` through `firewalld` in order to ensure that the other servers are able to access the block storage without issue. We will also reload the firewall, or the opening you just made will not be there:

```
$ sudo firewall-cmd --permanent --add-service=iscsi-target
$ sudo firewall-cmd -reload
```

After that, we will be utilizing the new service we just installed—`targetcli`—to create network block storage in order to share it with `rhel2.example.com`:

```
[emcleroy@rhel1 ~]$ sudo targetcli
targetcli shell version 2.1.53
Copyright 2011-2013 by Datera, Inc and others.
For help on commands, type 'help'.

/> ls
o- / ..................................................................... [...]
  o- backstores .......................................................... [...]
  | o- block ............................................... [Storage Objects: 0]
  | o- fileio .............................................. [Storage Objects: 0]
  | o- pscsi ............................................... [Storage Objects: 0]
  | o- ramdisk ............................................. [Storage Objects: 0]
  o- iscsi ........................................................ [Targets: 0]
  o- loopback ..................................................... [Targets: 0]
/>
```

Figure 1.14 – targetcli initiated for the first time

We will now create backstores for the physical disk partitioning. We will be creating backstores of the type `block` in order to meet the needs of the iSCSI system today. This will allow the persistent filesystems and us to set up how we would like the LUNs to be in terms of size and security:

```
/> cd /backstores/block
/backstores/block> create blockstorage1 /dev/sdb
Created block storage object blockstorage1 using /dev/sdb.
```

Next, we will create an IQN in the `/iscsi` directory in order to provide a target and destination for the block storage device:

```
/backstores/block> cd /iscsi
/iscsi> create iqn.2022-05.com.example:rhel1
Created target iqn.2022-05.com.example:rhel1.
Created TPG 1.
Global pref auto_add_default_portal=true
Created default portal listening on all IPs (0.0.0.0), port
3260.
/iscsi> ls
o- iscsi ................................... [Targets: 1]
  o- iqn.2022-05.com.example:rhel1 .............. [TPGs: 1]
    o- tpg1 ........................ [no-gen-acls, no-auth]
      o- acls .................................... [ACLs: 0]
      o- luns .................................... [LUNs: 0]
      o- portals .............................. [Portals: 1]
        o- 0.0.0.0:3260 ............................... [OK]
```

As you can see in the preceding code snippet, a default portal was created on port 3260 for connectivity to the block storage backstores using the `create` command for the IQN. Next, we will create a LUN for physically supporting the storage needs of the iSCSI block storage:

```
/iscsi> cd /iscsi/iqn.2022-05.com.example:rhel1/tpg1/luns
/iscsi/iqn.20...sk1/tpg1/luns> create  /backstores/block/
blockstorage1
Created LUN 0.
```

The next thing we need for iSCSI is an ACL to allow traffic to reach our storage devices successfully. We will need to exit out of `targetcli` temporarily to view the Red Hat name for the initiator's IQN. It can be found in `/etc/iscsi/initiatorname.iscsi`:

```
Global pref auto_save_on_exit=true
Configuration saved to /etc/
[emcleroy@rhel1 ~]$ vi /etc/iscsi/initiatorname.iscsi
```

Here is an example of the initiator name that is currently being used on the next image:

Figure 1.15 – initiatorname.iscsi

We will go back into `targetcli` and finish up the system preparations, setting up the system to use an ACL of our choosing for the system that will be utilizing the block storage:

```
[emcleroy@rhel1 ~]$ sudo targetcli
targetcli shell version 2.1.53
Copyright 2011-2013 by Datera, Inc and others.
```

```
For help on commands, type 'help'.

> cd /iscsi/iqn.2022-05.com.example:rhel1/tpg1/acls

/iscsi/iqn.20...sk1/tpg1/acls> create iqn.2022-05.com.
example:rhel2
Created Node ACL for iqn.2022-05.com.example:rhel2
Created mapped LUN 0.
```

Next, we will remove the default portal and create one on the specific IP address of our server:

```
> cd /iscsi/iqn.2022-05.com.example:rhel1/tpg1/portals

/iscsi/iqn.20.../tpg1/portals> delete 0.0.0.0 3260
Deleted network portal 0.0.0.0:3260
/iscsi/iqn.20.../tpg1/portals> create 192.168.1.198 3260
Using default IP port 3260
Created network portal 192.168.1.198:3260.
```

Finally, the following is your completed block storage target:

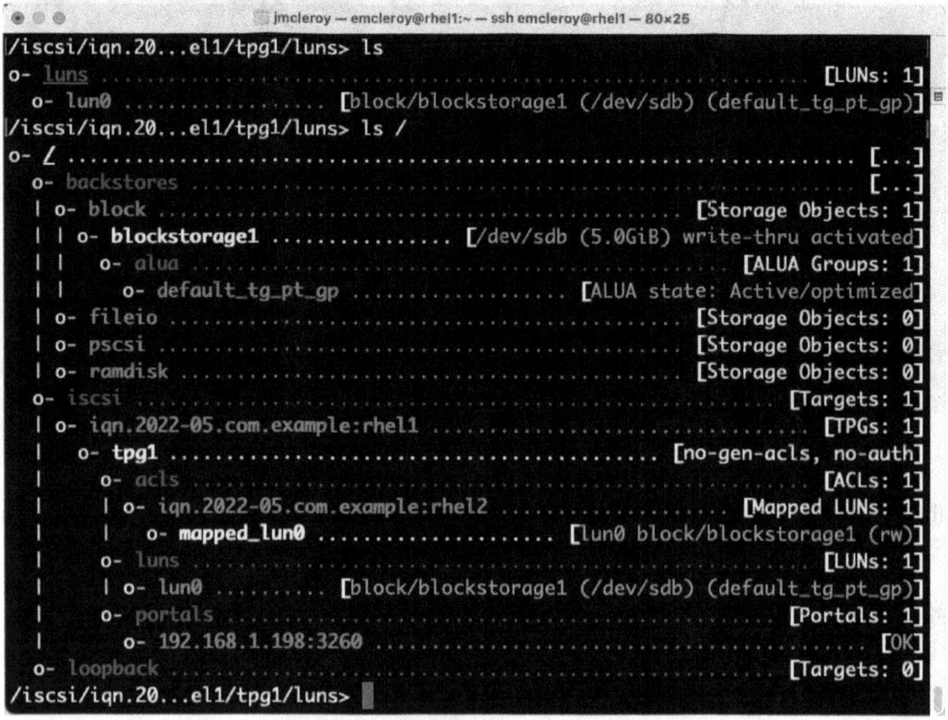

Figure 1.16 – iSCSI block storage target

We have just shown how to provision iSCSI block storage for consumption. Now, we will showcase how to utilize that block storage for actual usage in your systems. We will connect from `rhel1.example.com` to `rhel2.example.com` and mount it, provision it, and utilize it to move and store files, as one of the examples of how we can use these systems is to increase the storage capacity of remote servers without needing to increase space, power, or cooling directly for the rack the server is housed within.

The first thing we will need to do is install the iSCSI utilities, as on the exam you may not have the installation of the **Server with the GUI**:

```
$ sudo dnf install iscsi-initiator-utils targetcli -y
```

This allows us to ingest the iSCSI block storage that we created previously. Next, we are going to look up the configured target on `rhel1` (`192.168.1.198`) (please note: this might be a different IP for you in your lab) and log in to it to ensure connectivity. From here, we need to set the login information on the `/etc/iscsi/iscsid.conf` file in order to pass the correct login information so that we can log in to the storage device:

```
$ sudo getent hosts rhel1
```

Now, we will set the `InitiatorName` field so that we can pass a known entry to the connecting server using the following commands:

```
[emcleroy@rhel1 ~]$ sudo vi /etc/iscsi/initiatorname.iscsi
InitiatorName=iqn.2022-05.com.example:rhel1
[emcleroy@rhel1 ~]$ sudo systemctl restart iscsid.service
```

Please note you can use the manual page to gain further insight into the `iscsiadm` command set with the `man iscsiadm` command. On `rhel2`, we will do a discovery of available block devices using the `iscsiadm` command. The `-m` flag specifies the mode—in this case, `discovery`. The `-t` flag specifies the type of target—in our case, `st`, which is `sendtargets`, which tells the server to send a list of iSCSI targets. The `-p` flag specifies which portal to use, which is a combination of IP address and port. If no port is passed, it will default to `3260`:

```
[emcleroy@rhel2 ~]$ sudo iscsiadm -m discovery -t st -p
192.168.1.198:3260
```

Please note the output from the preceding command will be as follows:

```
192.168.1.198:3260,1 iqn.2022-05.com.example:rhel1
```

As you can see here, we have a block device that is showing as available.

We will try to log in to the device, and you can see we have logged in and it is showing the device connected, as follows:

```
[emcleroy@rhel2 ~]$ sudo iscsiadm -m node -T iqn.2022-05.com.
example:rhel1  -p 192.168.1.198 -l
```

In the preceding code, we are using the -m flag to choose node mode. We are using the -T flag to specify the target name. We are again using the -p flag for the portal, which defaults to port 3260. Finally, we are using the -l flag to tell iscsiadm to log in to the target.

Next, we are going to use the -m mode flag to check the session and -P to print the information level of 3:

```
[emcleroy@rhel2 ~]$ sudo iscsiadm -m session -P 3
iSCSI Transport Class version 2.0-870
version 6.2.1.4-1
Target: iqn.2022-05.com.example:rhel1 (non-flash)
     Current Portal: 192.168.1.198:3260,1
     Persistent Portal: 192.168.1.198:3260,1
```

You can see that we have sdb, which is the second drive on rhel2, and now we have sdc as well:

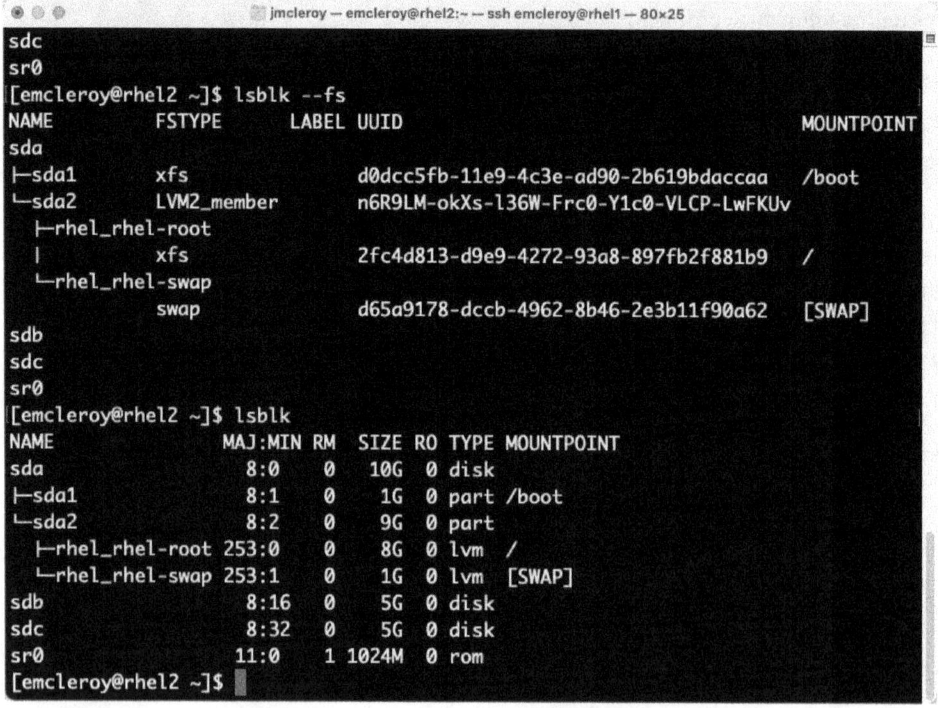

Figure 1.17 – sdc drive is now showing up

Next, we are going to partition the drive and format it with xfs. This will allow us to mount the system on boot as well as to save persistent files. This can be used for many things from file storage to OS or databases. First, we are going to format the drive to xfs:

```
[emcleroy@rhel2 ~]$ sudo mkfs.xfs /dev/sdc
meta-data=/dev/sdc              isize=512      agcount=4,
agsize=327680 blks
         =                      sectsz=512     attr=2,
projid32bit=1
         =                      crc=1          finobt=1,
sparse=1, rmapbt=0
         =                      reflink=1
data     =                      bsize=4096     blocks=1310720,
imaxpct=25
         =                      sunit=0        swidth=0 blks
naming   =version 2             bsize=4096     ascii-ci=0,
ftype=1
log      =internal log          bsize=4096     blocks=2560,
version=2
         =                      sectsz=512     sunit=0 blks,
lazy-count=1
realtime =none                  extsz=4096     blocks=0,
rtextents=0
```

Then, we are going to use the following command to get the UUID to use in fstab to make it a persistent mount that will automatically mount at startup:

```
[emcleroy@rhel2 ~]$ lsblk -f /dev/sdc
NAME FSTYPE LABEL
UUID                                    MOUNTPOINT
sdc  xfs            38505868-00de-4269-88d8-3357a22f2101
[emcleroy@rhel2 ~]$ sudo vi /etc/fstab
```

Here, we can see an example of the added value highlighted in fstab:

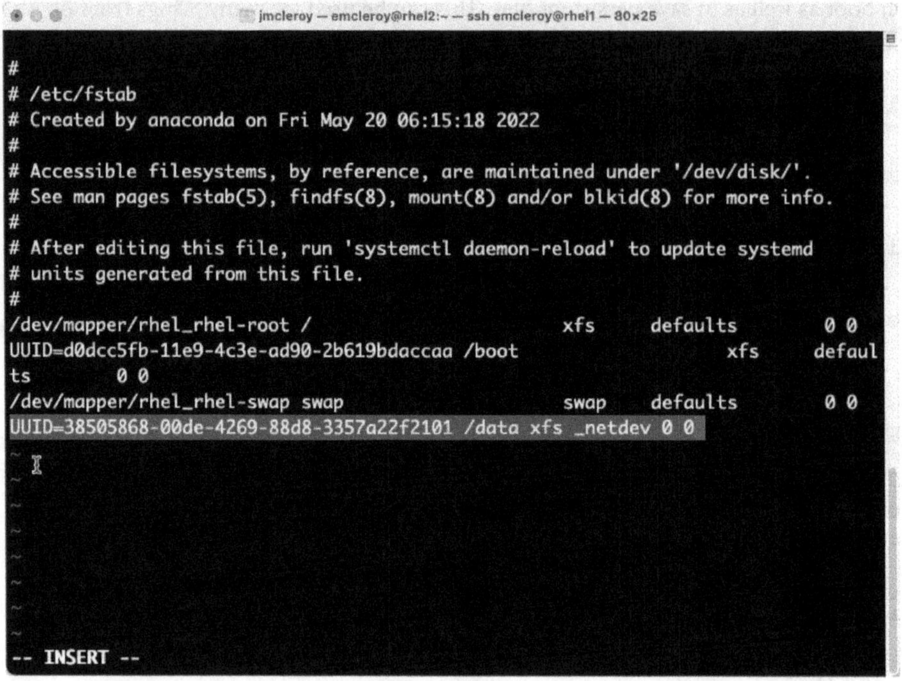

Figure 1.18 – Updated fstab after adding the iSCSI block storage device

Here are the lines where we added the new iSCSI drive to fstab. Please note that for network devices, we pass the _netdev option. Next, we are going to mount the system in order to use it for moving files around:

```
[emcleroy@rhel2 ~]$ sudo mkdir -p /data
[emcleroy@rhel2 ~]$ sudo mount /data
[emcleroy@rhel2 ~]$ df /data
Filesystem       1K-blocks   Used  Available Use% Mounted on
/dev/sdc          5232640 69616    5163024    2% /home/emcleroy/
data
[emcleroy@rhel2 ~]$ cd /data
```

After it is mounted, we are going to move into the new drive, create a folder and a test .txt file, and ensure it saves, which it does by using the following commands:

```
[emcleroy@rhel2 ~]$ sudo mkdir test
[emcleroy@rhel2 ~]$ cd test/
[emcleroy@rhel2 ~]$ sudo vi test.txt
```

Next, we are going to remove the mount, log out of the connection, and delete the leftovers:

```
[emcleroy@rhel2 ~]$ cd
[emcleroy@rhel2 ~]$ sudo umount /data
[emcleroy@rhel2 ~]$ sudo iscsiadm -m node -T iqn.2022-05.com.
example:rhel1 -p 192.168.1.198 -u
Logging out of session [sid: 8, target: iqn.2022-05.com.
example:rhel1, portal: 192.168.1.198,3260]
Logout of [sid: 8, target: iqn.2022-05.com.example:rhel1,
portal: 192.168.1.198,3260] successful.
[emcleroy@rhel2 ~]$ sudo iscsiadm -m node -T iqn.2022-05.com.
example:rhel1 -p 192.168.1.198 -o delete
```

This wraps up the section on manually setting up iSCSI. Next is automating it. We will go into more detail in the hands-on review and the quiz at the end of the book. I hope you are enjoying this journey as much as I am.

iSCSI block storage – Ansible automation playbook creation and usage

We will start the automation portion of working with iSCSI block storage by first installing and configuring the use of Ansible core 2.9 as that is what is used in the *EX358* exam. I will not be using the **fully qualified collection name** (**FQCN**) as that can sometimes cause errors in a 2.9 environment, which could lead to issues while taking the exam. This we want to avoid at all costs, so we will be using the classic module names, and I will explain the differences to a degree so that you can understand what you will need to use in future versions of Ansible.

First, let's start by installing Ansible 2.9 on server rhel3 as that is going to be what we consider the workstation server from our yum repository. Depending on your personal preferences, you can make rhel1 your classroom server and rhel2 and rhel3 your test servers, but in our case, we have already set up rhel1 with iSCSI and rhel2.

First, we will enable the needed repos:

```
[emcleroy@rhel3 ~]$ sudo subscription-manager repos --enable
ansible-2.9-for-rhel-8-x86_64-rpms
Repository 'ansible-2.9-for-rhel-8-x86_64-rpms' is enabled for
this system.
```

Next, we will install Python 3:

```
[emcleroy@rhel3 ~]$ sudo dnf install python3 -y
```

Then, we will install Ansible 2.9:

```
[emcleroy@rhel3 ~]$ sudo dnf install ansible -y
```

Let's check and ensure that the right version of Ansible is installed:

```
[emcleroy@rhel3 ~]$ ansible --version
ansible 2.9.27
  config file = /etc/ansible/ansible.cfg
  configured module search path = ['/home/emcleroy/.ansible/
plugins/modules', '/usr/share/ansible/plugins/modules']
  ansible python module location = /usr/lib/python3.6/site-
packages/ansible
  executable location = /usr/bin/ansible
  python version = 3.6.8 (default, Oct 11 2019, 15:04:54) [GCC
8.3.1 20190507 (Red Hat 8.3.1-4)]
```

Next, we are going to start writing a playbook using the **Yet Another Markup Language** (**YAML**) Ansible language. This is a simple module-based function that will allow you to write up a playbook that will accomplish your task quickly and efficiently. I recommend a good editor when writing up these playbooks. JetBrains' PyCharm is my go-to and is what you will see me write my playbooks in when you see example screenshots of the finished results. Do also note that the finished playbooks can be found in the GitHub repository of this book, as mentioned in the *Technical requirements* section for each chapter.

The first thing you will want to create is a directory where you want to run the playbooks from:

```
[emcleroy@rhel3 ~]$ mkdir iscsi_mount
```

Once in the directory, we will create an inventory file with a default group that will have both the rhel1 and rhel2 servers in them:

```
[emcleroy@rhel3 ~]$ cd iscsi_mount
[emcleroy@rhel3 ~]$ vi inventory
[defaults]
rhel1 ansible_host=192.168.1.198
rhel2 ansible_host=192.168.1.133

[iscsi_block]
rhel1 ansible_host=192.168.1.198
```

```
[iscsi_user]
rhel2 ansible_host=192.168.1.133
```

As you can see, I added `ansible_host` and the IP address. This is in case there is no host file set up or the name is not DNS routable. I added the default group with all of the hosts, and there are two additional groups that allow me to limit what my playbooks make changes to. That way, I can tell my playbook to mount the storage on `rhel2` using the `iscsi_user` group.

Next, we are going to write the block storage playbook named `mount_iscsi.yml`, and I will break it down after showing you what that playbook looks like:

```
---
- name: Ensure /data is mounted from rhel1 iSCSI target that
  was created manually onto rhel2
  hosts: iscsi_user
  become: true
  become_method: sudo
  tasks:
    - name: the targetcli package is installed
      yum:
        name: targetcli
        state: present
    - name: the IQN is set for the initiator
      template:
        dest: /etc/iscsi/initiatorname.iscsi
        src: templates/initiatorname.iscsi.j2
        mode: '644'
        owner: root
        group: root
    - name: Create mount directory for /data
      file:
        path: /data
        state: directory
        mode: '0755'
    - name: Restart iscsiadm
      command:
        cmd: systemctl restart iscsid.service
    - name: Mount new drive
      command:
```

```
cmd: iscsiadm -m node -T iqn.2022-05.com.example:rhel1   -p
192.168.1.198 -l
```

The module name for this instance is `yum`, and that is used to install the `iscsi-initiator-utils` package that will install the utilities. Next, we have the different flags of the modules, such as `dest:` for the destination of the source file that is in your playbook's `templates` folder. In the template folder location within your playbook directory, you will have the file `/templates/initiatorname.iscsi.j2`, which contains the initiator name to pass to the playbook. It will contain the following code:

```
InitiatorName=iqn.2022-05.com.example:rhel1
```

You can find out more about each module that you're using by looking at the equivalent of a man page, as follows:

```
[emcleroy@rhel3 ~]$ ansible-doc yum
```

You can also list the files with the following command, but keep in mind there are thousands of modules, so try to grep the names if possible:

```
[emcleroy@rhel3 ~]$ ansible-doc --list
```

The following screenshot shows what a normal `ansible-doc` page looks like for the different modules:

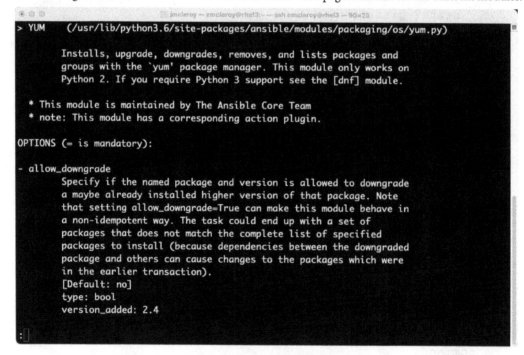

Figure 1.19 – Example of the yum module documentation page

We will use the following command to run the `ansible-playbook -i inventory mount_iscsi.yml -u emcleroy -k --ask-become -v` playbook. This will be executed from the `rhel3` server and make changes to the `rhel2` server. This concludes our automated approach to mounting a LUN for iSCSI block storage. We learned a little about Ansible and how it works, from modules to templates. We will learn a lot more about Ansible and all of its inner workings in the upcoming chapters in greater detail, so stick around.

Summary

This brings us to the end of the first chapter, where we went into details about RHEL block storage, setting up a hands-on environment for testing purposes, and getting the first taste of Ansible. In the coming chapters, we will be digging deeper into how to use Ansible with further examples and more hands-on exercises that will help hone your abilities as a systems admin and help ensure you pass the *EX358* exam. In the next chapter, we will be continuing our journey into network storage, talking about network file storage and how we can use that to share information across our organizations and make our jobs faster. Please join me as we continue our road to gaining the *EX358* certification that you want to achieve and that I want to help you obtain.

2

Network File Storage – Expanding Your Knowledge of How to Share Data

In this chapter, we're going to cover the topics of **Network File Storage (NFS)** and Samba using **Server Message Block (SMB)** storage. These are two network-based file-sharing options. One (NFS) is heavily used in Linux, while the other is the mainstay of Windows, which is SMB (in our case, Samba). We are going to learn how to set up both of these target folders and share them with the world. We will go over making sure they are only available to those we are attempting to share access with. We will show how we can ensure that unauthorized access does not occur – well, to the best of our abilities; people are sneaky.

After we set those up manually, we will work on setting them up through automation using Ansible. This can make it extremely easy to set up new servers to access file shares that are needed for every new server in an automated fashion. This ensures that each new server has the required access without worrying whether you forgot to add one of the folders. Automation helps take the headache out of system admin work and with Ansible, it is even easier. Let's get started in the world of NFS; it only makes our lives easier.

In this chapter, we will cover the following topics:

- NFS and SMB network storage – the way they work and when to choose one over the other
- NFS manual provisioning and deployment
- NFS Ansible Automation playbook creation and usage
- SMB storage manual provisioning and deployment
- SMB storage Ansible Automation playbook creation and usage

Technical requirements

This section covers the technical requirements for this chapter.

Setting up GitHub access

Please refer to *Chapter 1, Block Storage – Learning How to Provision Block Storage on Red Hat Enterprise Linux*, for gaining access to GitHub. You will find the Ansible Automation playbooks for this chapter at the following link: `https://github.com/PacktPublishing/Red-Hat-Certified-Specialist-in-Services-Management-and-Automation-EX358-Exam-Guide/tree/main/Chapter02`. Remember, these are suggested playbooks and are not the only way you can write them to make the playbooks work for you.

You can always change them up using `raw`, `shell`, or `cmd` to achieve the same results, but we are demonstrating the best way to accomplish the goals. Also, keep in mind that we are not using the **Fully Qaulified Collection Name (FQCN)** that is needed in the future version of Ansible, as that will not be supported in the exam as it is testing against Ansible 2.9.

Setting up your lab environment for NFS and SMB

Please refer to the setup of the lab environment from *Chapter 1, Block Storage – Learning How to Provision Block Storage on Red Hat Enterprise Linux*, if not already completed. You can also use this opportunity to create snapshots of your environment so that you can revert to them and do the hands-on exercises over and over again. This can be accomplished as follows.

Snapshots are easy to create in VirtualBox: you simply need to click on the following button that is selected with the mouse cursor to create a snapshot:

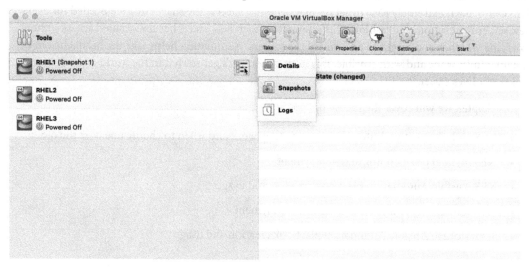

Figure 2.1 – VirtualBox options button to open snapshot creation

Click to take one at the top and then name your snapshot. This will allow you to determine the state of the machine image when the snapshot was taken. This can be seen in the following figure:

Figure 2.2 – VirtualBox snapshot save feature

This will wrap up taking a snapshot in order to revert to previous states for repeated hands-on rework. Now, let's get into the storage wars.

NFS and SMB network storage – the way they work and when to choose one over the other

NFS is a great way to share files across Linux systems. It is machine-based authentication so the system can have access to the files for all users. This is great for Linux when sharing small files as the speed is increased and the system shares are retrieved more efficiently. SMB, on the other hand, is user authentication-based and better suited for interactions with Windows computers. So, if you have a multivendor environment, it might be worthwhile to set up SMB as it has the ability to share printers as well. This allows for file sharing as well as printer sharing, thus giving a slight multi-functional edge to the system to make up for the slight speed loss.

The differences in authentication make for some differences in setup and can lead to some headaches when it comes to making sure you are locking down the right permissions. SMB using user-based authentication is a bit better because, with NFS, anyone with access to the allowed machine can gain access to the files being shared. So, in reality, it is more of a toss-up of what you need, such as security or multivendor support in your environment, when it comes to choosing to implement either of the two. In most cases, both are actually implemented for different system uses.

Let's go into a little more depth on NFS and what it can do for RHEL 8.1 users and Linux overall through setting it up hands-on. This will show you how to set up NFS manually and through automation. RHEL 8.1 supports NFSv3 and NFSv4. NFSv3 can use UDP and TCP, whereas NFSv4 uses TCP. Older versions are no longer supported. Let's set up NFS on our systems.

Initially, we are going to install the package needed if it is not already installed. This package is the nfs-utils package and is needed for the client and the target machines. It provides all the tooling to access folders and mount the needed folders. By installing it on both the server and client, you ensure that you are able to enable all of the dependencies that are needed in order to run the service.

Let's install nfs-utils as shown in the following figure:

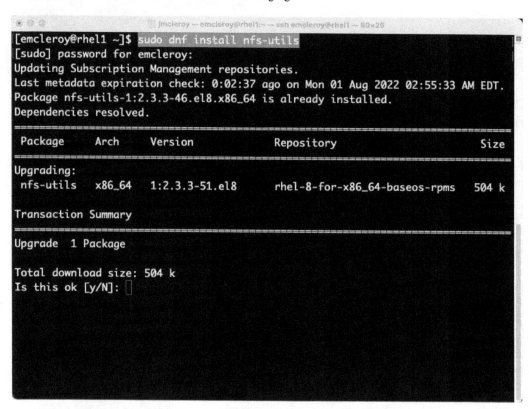

Figure 2.3 – Installing the nfs-utils package

Now, there is a small caveat that can lead to frustration – trust me, when writing this and running into the issue, I was boiling! After a large amount of troubleshooting, the current 2.3.3-51.el8 has a bug that will not allow the exports to share out correctly and the proper permissions to connect. So, after installing NFS, we need to downgrade it one step to 2.3.3-46.el8, where the bug is not present. In the following figure, it is shown being downgraded:

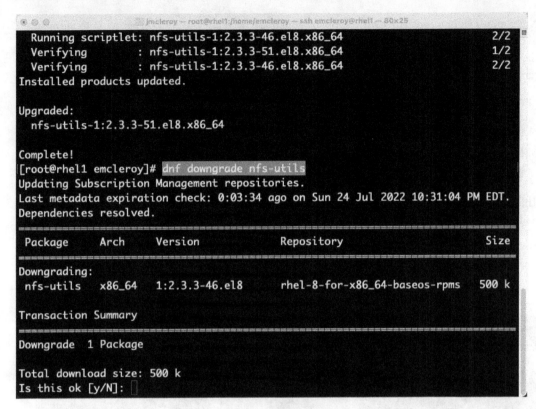

```
    Running scriptlet: nfs-utils-1:2.3.3-46.el8.x86_64                2/2
    Verifying        : nfs-utils-1:2.3.3-51.el8.x86_64                1/2
    Verifying        : nfs-utils-1:2.3.3-46.el8.x86_64                2/2
Installed products updated.

Upgraded:
  nfs-utils-1:2.3.3-51.el8.x86_64


Complete!
[root@rhel1 emcleroy]# dnf downgrade nfs-utils
Updating Subscription Management repositories.
Last metadata expiration check: 0:03:34 ago on Sun 24 Jul 2022 10:31:04 PM EDT.
Dependencies resolved.
================================================================================
 Package      Arch       Version          Repository                     Size
================================================================================
Downgrading:
 nfs-utils    x86_64     1:2.3.3-46.el8    rhel-8-for-x86_64-baseos-rpms   500 k

Transaction Summary
================================================================================
Downgrade   1 Package

Total download size: 500 k
Is this ok [y/N]: 
```

Figure 2.4 – Downgrading nfs-utils

This allows us to do our hands-on exercises without issue. Before, if you tried to do them, you would be banging your head on your keyboard much like I was.

After the correct version is installed, we are going to enable the daemon:

```
[emcleroy@rhel1 ~]$ sudo systemctl enable --now nfs-server
Created symlink /etc/systemd/system/multi-user.target.wants/
nfs-server.service → /usr/lib/systemd/system/nfs-server.
service.
```

Then, we have to open the firewall to ensure that connectivity is allowed:

```
[emcleroy@rhel1 ~]$ sudo firewall-cmd --permanent --add-
service=nfs
success
[emcleroy@rhel1 ~]$ sudo firewall-cmd --reload
Success
```

Next up, we will be configuring the exports, which are what tell the NFS server what to allow out to the world. The settings are granular and allow for thorough control. The exports allow you to tell the system what can be read and by what machines. You can allow specific machines via DNS, IP, IP range, and so on.

First, let's make a shared directory as root:

```
[emcleroy@rhel1 ~]$ sudo su -
[root@rhel1 ~]# cd
[root@rhel1 ~]# mkdir -p /share/folder
```

This will be the folder we share out to the world. Others will access this to read and write files. We will be editing the /etc/exports file to share out this folder.

Here are some examples of a correctly formatted exports file. We will talk more about the settings shown in the following screenshot, such as (rw):

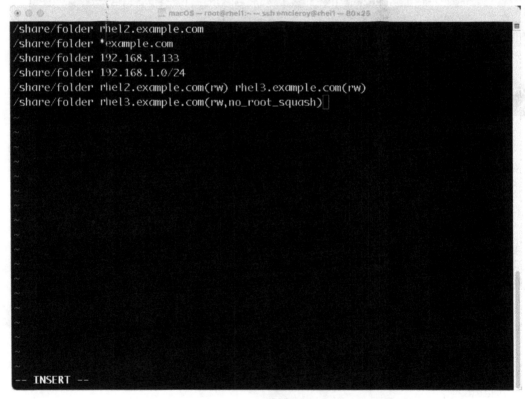

Figure 2.5 – /etc/exports example file for NFS share folder

As you can see, you can export NFS shares via DNS as noted. You can export via IP and IP range. You can use space delimited for multiple shared locations if you do not wish to stack them. You can also put restraints such as read-only or read-write. The no_root_squash function allows for root access to the directory instead of the nobody user being selected. By default, if a root client creates a file, the nobody user owns that file. This can cause issues with access to the file, so it is a good idea to use no_root_squash. This gives full root access to the files. Please be aware that the preceding is an example, and duplicates such as the ones shown will be ignored by NFS.

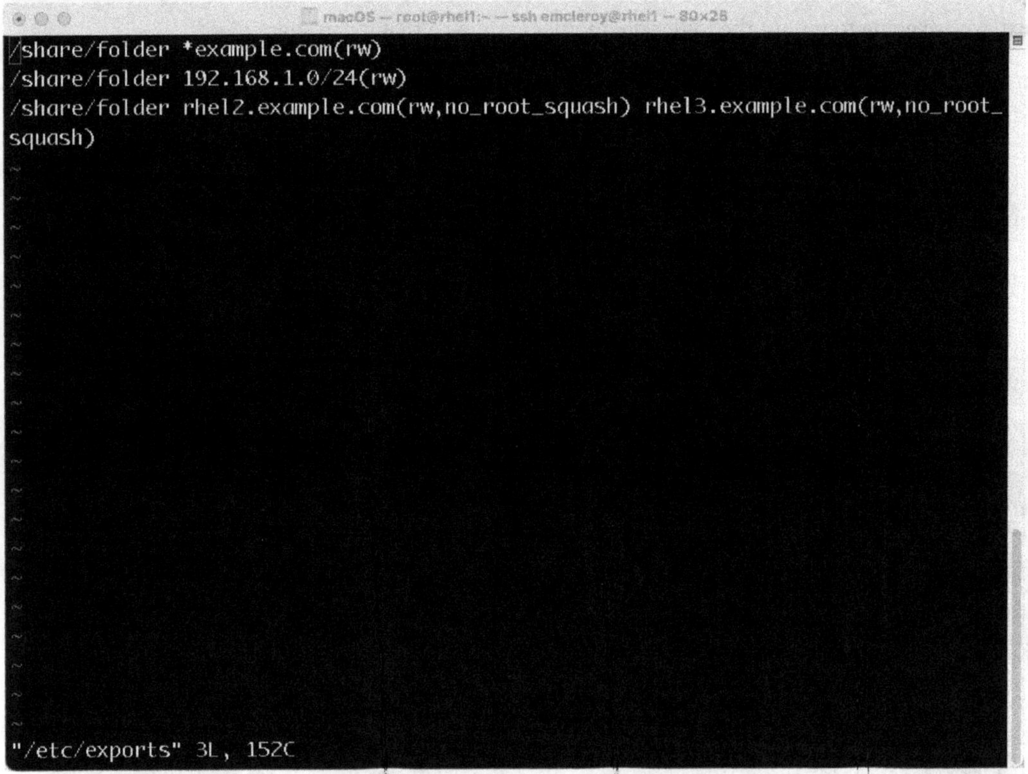

```
/share/folder *example.com(rw)
/share/folder 192.168.1.0/24(rw)
/share/folder rhel2.example.com(rw,no_root_squash) rhel3.example.com(rw,no_root_
squash)
~
~
~
~
~
~
~
~
~
~
~
~
~
~
~
~
~
~
~
~
"/etc/exports" 3L, 152C
```

Figure 2.6 – Correct NFS /etc/exports file

You then need to run exportfs -r in order to update the exports so that they are served to the endpoints:

```
[root@rhel1 ~]# exportfs -r
[root@rhel1 ~]#
```

You can then check your exports to ensure they are being served with the `exportfs` command:

```
[root@rhel1 ~]# exportfs
/share/folder    rhel2
/share/folder    rhel3
/share/folder    192.168.1.0/24
/share/folder    *example.com
```

Next, we are going to mount the folder on another system so that it can share files.

On RHEL3, we are going to run the `mount` command to mount the filesystem. You can also add this to `fstab` to make this permanent:

```
[root@rhel3 ~]# mkdir -p /share
[root@rhel3 ~]# mount -t nfs 192.168.1.198:/share/folder /share
```

Here is an example of `fstab` for NFS as well:

```
jmcleroy — root@rhel3:~ — ssh emcleroy@rhel3 — 80×25

#
# /etc/fstab
# Created by anaconda on Fri May 20 06:15:18 2022
#
# Accessible filesystems, by reference, are maintained under '/dev/disk/'.
# See man pages fstab(5), findfs(8), mount(8) and/or blkid(8) for more info.
#
# After editing this file, run 'systemctl daemon-reload' to update systemd
# units generated from this file.
#
/dev/mapper/rhel_rhel-root /                          xfs      defaults       0 0
UUID=d0dcc5fb-11e9-4c3e-ad90-2b619bdaccaa /boot                   xfs      defaul
ts        0 0
/dev/mapper/rhel_rhel-swap swap                       swap     defaults       0 0
192.168.1.198:/share/folder /share                    nfs      defaults       0 0
~
~
~
~
~
~
-- INSERT --
```

Figure 2.7 – Example of what fstab would look like

After that, you should be able to go to the folder and add files to share. The other systems with access can also see these files.

This concludes the walk-through of NFS and how to export it to systems manually. This is what you will be tested on in the exam. The bonus mounting of the system is for your understanding, and I hope that helps you in the future. Now we are going to talk about the automation of the process of NFS server provisioning.

NFS Ansible Automation playbook creation and usage

We are going to delve deeper into the Ansible Automation playbook. We will fully break it down for the NFS exports. We are going to show each step of the playbook and what each item does. By *item*, we are referring to the variables and commands set within an Ansible playbook:

```
---
- hosts: exports
  become: true
  become_method: sudo
```

Here, we have the beginning of the playbook. The hosts are the targets of the playbook. These are normally the servers you want to make changes to. These are listed in the inventory file. In this case, we are targeting the exports group. In this case, become tells Ansible that you want to elevate the user, and become_method tells it that you want to use sudo to elevate the permissions:

```
  tasks:
  - name: Make sure Directory exists
    file:
      path: /share/folder
      owner: root
      state: directory
```

We are now going to move to tasks. These are the actions we want Ansible to take. The first task we see here is to create a directory using the file module. The name field is like a comment for a better understanding of the code by other engineers. From there, we see the other items we set, and then there are assumed defaults. As mentioned in *Chapter 1, Block Storage – Learning How to Provision Block Storage on Red Hat Enterprise Linux*, you can use the ansible-doc file to see the usage and examples:

```
  - name: Install nfs utilities
    package:
      name: nfs-utils
      state: "2.3.3-46.el8"
```

Normally, you would use the latest state; however, in our case, that would cause the system to encounter the bug that fails connectivity. It should work fine with the latest on the exam though.

In this task, we installed the `nfs-utilities` package using the generic package installer for Ansible. This uses the facts gathered by Ansible. By using those facts, it knows to use `dnf` to install the software:

```
- name: the directory is shared
  copy:
    content: "/share/folder *(rw,no_root_squash)\n"
    dest: /etc/exports
```

Using the `copy` module, we are replacing the current `exports` file. If this was a multiline exports file, it would need to be a template, `lineinfile`, or all lines would need to be in the task. Here, we are using the `service` module to enable and start the `nfs-server` service. This will ensure that the service is started. The service is enabled upon boot. The service also loads the exports, as `nfs-server` is started after the `exports` file was updated:

```
- name: Enable and Start NFS Service
  service:
    name: nfs-server
    state: started
    enabled: yes
```

Using the `firewalld` module, we are going to allow a permanent opening for the NFS file service. This will allow connectivity to the resources we opt to share. Once that is complete, the `exports` playbook is completed, and up next is the `clients` playbook:

```
- name: Open firewall for NFS-Server
  firewalld:
    service: nfs
    immediate: yes
    permanent: yes
    state: enabled
```

After we have finished creating the playbook, we will use the `ansible-playbook -i inventory nfs_server.yml -u emcleroy -k --ask-become -v` command to execute the playbook. The flag of `-i` indicates the name of the inventory to use. The `-u` flag indicates the username to use when executing the playbook. The `-k` flag indicates to ask for the user password. The `--ask-become` flag specifies that you want to provide the `sudo` password for the playbook run. Finally, `-v` shows additional verbosity so you can see more logs of what is happening when the playbook is running.

Here, we will show a quick view of the playbook used to support the client setup of utilizing a share that has been set up for us:

```
---
- hosts: clients
  become: true
  become_method: sudo
  tasks:
  - name: Install nfs utilities
    package:
      name: nfs-utils
      state: present

  - name: Add the NFS share and mount it with fstab
    mount:
      path: /share
      src: 192.168.1.198:/share/folder
      state: mounted
      fstype: nfs
```

This will install `nfs-utilities` and then mount the file to the `/share folder`.

To run the playbook, we will use the following command: `ansible-playbook -i inventory nfs_client.yml -u emcleroy -k --ask-become -v`. That wraps up our foray into NFS. We will now be moving on to some more network-based file storage. We will be discussing SMB, which is most commonly used for Windows systems. We will go through setting it up manually and with automation.

SMB storage manual provisioning and deployment

The major difference when it comes to SMB is that it is the sharing standard for Windows-based systems from Linux systems. The best way to learn about Samba (another name for SMB) is to get your hands dirty, and that is what we are going to do today. We will be going step by step into how to install and provision SMB.

To start, we are going to install SMB via `dnf`:

```
[emcleroy@rhel1 ~]$ sudo dnf install samba -y
```

Then, create the share for SMB to serve:

```
[emcleroy@rhel1 ~]$ sudo mkdir -p /share/smbfolder
```

Next, we are going to start and enable the `smb` service:

```
[emcleroy@rhel1 ~]$ sudo systemctl enable --now smb
Created symlink /etc/systemd/system/multi-user.target.wants/
smb.service → /usr/lib/systemd/system/smb.service.
```

After that, we are going to open the firewall service for Samba:

```
[emcleroy@rhel1 ~]$ sudo firewall-cmd --permanent --add-
service=samba
success
[emcleroy@rhel1 ~]$ sudo firewall-cmd  --reload
success
```

With SMB now installed and the firewall opened, we can move on to setting up the folder we created. This will allow us to set the SELinux permissions needed. We are going to give access to administrators:

```
[emcleroy@rhel1 ~]$ sudo groupadd admins
[emcleroy@rhel1 ~]$ sudo chgrp admins /share/smbfolder/
```

This will allow us to restrict access to the folder to the `admins` group only.

With SELinux in the enforcing mode, it needs the correct context. As shown in these commands, the correct context is added back to the folder:

```
[emcleroy@rhel1 ~]$ sudo semanage fcontext -a -t samba_share_t
"/share/smbfolder(/.*)?"
[emcleroy@rhel1 ~]$ sudo restorecon -Rv /share/smbfolder
```

This allows for the label to stay, even during relabeling events within SELinux. SELinux context is outside the scope of this class and can be researched further online or through other Packt books.

Next, we will work on setting up the `smb.conf` file with the correct information to allow access to the `share` folder:

```
[emcleroy@rhel1 ~]$sudo vi /etc/samba/smb.conf
```

From there, you can see an example of a global Samba config portion and Samba share config portion in the following screenshot:

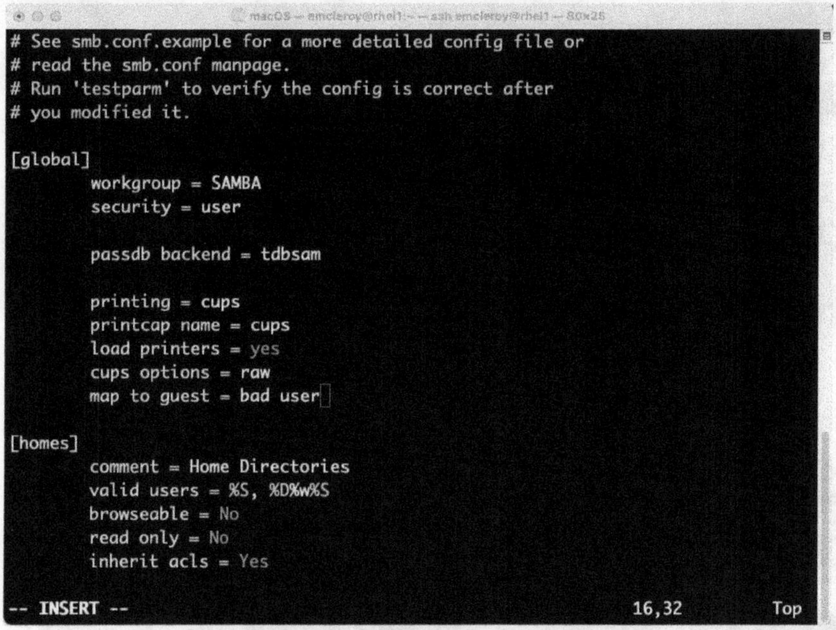

```
# See smb.conf.example for a more detailed config file or
# read the smb.conf manpage.
# Run 'testparm' to verify the config is correct after
# you modified it.

[global]
        workgroup = SAMBA
        security = user

        passdb backend = tdbsam

        printing = cups
        printcap name = cups
        load printers = yes
        cups options = raw
        map to guest = bad user

[homes]
        comment = Home Directories
        valid users = %S, %D%w%S
        browseable = No
        read only = No
        inherit acls = Yes

-- INSERT --                                    16,32        Top
```

Figure 2.8 – Global config for SMB with guests allowed

Next, we have the guests allowed setting for our share, as follows:

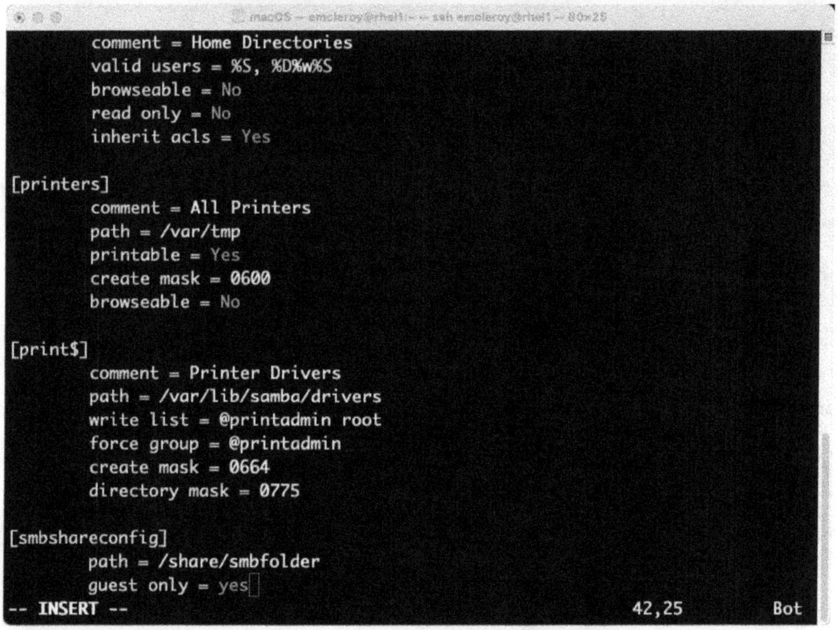

```
        comment = Home Directories
        valid users = %S, %D%w%S
        browseable = No
        read only = No
        inherit acls = Yes

[printers]
        comment = All Printers
        path = /var/tmp
        printable = Yes
        create mask = 0600
        browseable = No

[print$]
        comment = Printer Drivers
        path = /var/lib/samba/drivers
        write list = @printadmin root
        force group = @printadmin
        create mask = 0664
        directory mask = 0775

[smbshareconfig]
        path = /share/smbfolder
        guest only = yes
-- INSERT --                                    42,25        Bot
```

Figure 2.9 – Guests are allowed as noted in the guest only statement

Finally, here is an example of having to have a password to access the folder:

```
          comment = Home Directories
          valid users = %S, %D%w%S
          browseable = No
          read only = No
          inherit acls = Yes

[printers]
          comment = All Printers
          path = /var/tmp
          printable = Yes
          create mask = 0600
          browseable = No

[print$]
          comment = Printer Drivers
          path = /var/lib/samba/drivers
          write list = @printadmin root
          force group = @printadmin
          create mask = 0664
          directory mask = 0775

[smbshareconfig]
          path = /share/smbfolder
          guest only = no
-- INSERT --                                                    42,24          Bot
```

Figure 2.10 – As shown here, guest only is set to no so you need to have a user and password

These settings allow for control over your smb share.

Next, we are going to start the service and create some users. These users and passwords will only be good for the smb share. They will not affect the login information:

```
[emcleroy@rhel1 ~]$ sudo adduser -M sambauser -s /sbin/nologin
[emcleroy@rhel1 ~]$ sudo smbpasswd -a sambauser
New SMB password:
Retype new SMB password:
Added user sambauser.
```

Finally, we are going to start up smb so that we can use it for our newly created sambauser:

```
[emcleroy@rhel1 ~]$ sudo systemctl start smb
```

In the following screenshot, we can see that smb is running:

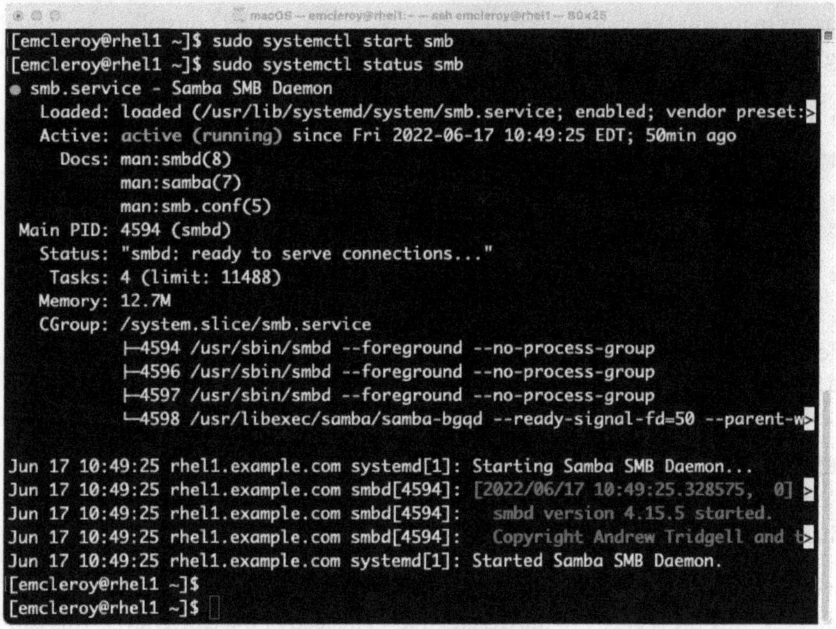

Figure 2.11 – SMB successfully running on the system

To mount the system, we will have to install `samba-client` on the client machine. As shown in the following screenshot, we are installing `samba-client` on `rhel2`:

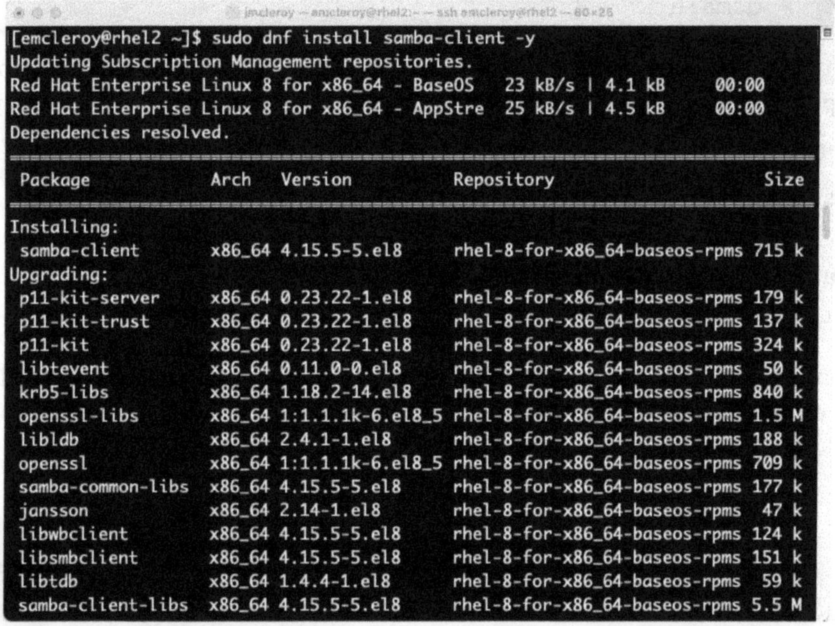

Figure 2.12 – samba-client being installed on rhel2

This will allow us to mount the share on `rhel2`. We need `cifs-utils` installed, and then we can mount using the following command:

```
[root@rhel2 ~]# dnf install cifs-utils
[root@rhel2 ~]# mkdir -p /share/folder
[root@rhel2 ~]# mount -t cifs -o username=sambauser
//192.168.1.198/smbshareconfig /share/folder
```

> **Note**
>
> This code and the process will all be available via the **Code in Action** (**CiA**) videos that are provided by Packt.

Next, we will go over setting up a restricted share to a client on a particular domain. Then we will show how to define which users/groups have read-only or write access. Finally, we will show how to mount permanently using specific user credentials.

Let's start with how to set up a restricted share to a client on a particular domain. This can be accomplished by setting the security to the `domain` level instead of the `user` level. This can be seen in the following code snippet:

```
[global]
workgroup = SAMBA
security = domain
```

Next, to set up a specific group to have read-write and also allow other users to have read-only, you would set up the smb file for the share to look something like the following when `security` is set to `user`:

```
[smbshareconfig]
path = /smb/smbfolder
valid users = systemadmin1, @engineers
write list = @engineers
```

The preceding file settings will allow `systemadmin1` to have read-only privileges and allow the `engineers` group to have read-write privileges, and all others will be unable to view the share.

Next, we are going to look at how to set up a permanent mount using CIFS credentials and `fstab` in order to mount the share at startup.

We will start by creating a CIFS credential file, as shown here:

```
[emcleroy@rhel1 ~]$ vi /tmp/creds.txt
```

This will have the following body of text:

```
username=smbuser
password=redhat
```

After the file is created, we will use it for login credentials for the mount and add the following to `fstab` to mount the share with these credentials:

```
//rhel1.example.com/smbshareconfig /mnt/shared cifs
credentials=/tmp/creds.txt,multiuser 0 0
```

That wraps up the manual portion of SMB storage. We will now work through the templates and other things needed for the automated approach of setting up SMB storage. Stay tuned for more great information about this and what we need to do to get it working through automation. This will help you with completing the EX358 exam as well.

SMB storage Ansible Automation playbook creation and usage

In this section, we will be looking at what is needed in order to set up SMB shares through Ansible Automation. This will allow for quicker setup across multiple servers. This will, in turn, also allow you to spend less time on the keyboard and more time focused on what you really want to learn. We have covered the different areas of Ansible in depth, from comments to names to modules. We are going to go through all of this again for SMB storage and show you how to set it up. This again is one way of doing it; there are others with Ansible that will be successful.

Let's get started as we have with all of the other playbooks we have created. First, we will choose the inventory hosts and escalation points that we want to use. I showed you how to set up the inventory in previous sections so we will skip that. However, here is a snapshot of the inventory for your reference:

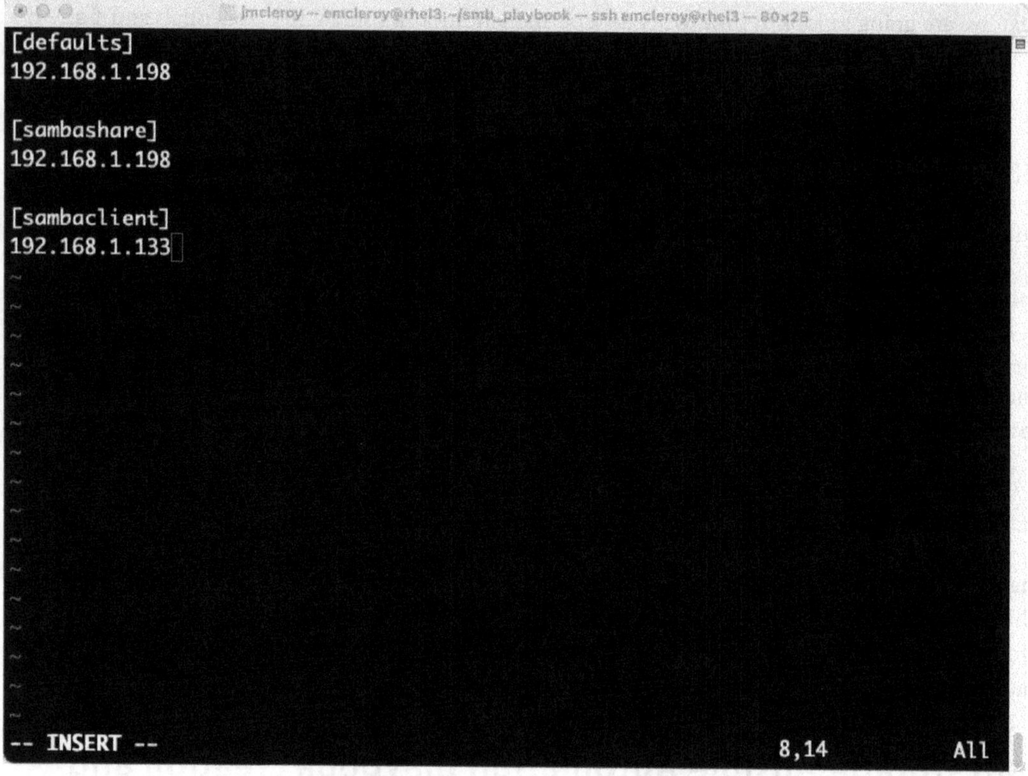

Figure 2.13 – smb_playbook inventory snapshot

To start, let's set up a new directory for the playbook:

```
[emcleroy@rhel3 ~]$ mkdir smb_playbook
[emcleroy@rhel3 ~]$ cd smb_playbook/
[emcleroy@rhel3 smb_playbook]$ vi smb_playbook.yml
---
- hosts: sambashare
  become: true
  become_method: sudo
```

Next, we are going to start the tasks to set up everything needed for samba. First, let's install it and the client to be thorough:

```
tasks:
  - name: Install Samba and Samba-Client
    package:
```

```
    name:
        - samba
        - samba-client
        - cifs-utils
    state: latest
```

This portion of the code creates the groups that we will use to set allow functions on SMB:

```
- name: Create groups that are allowed
  group:
    name: "admins"
    system: yes
```

This portion of the code creates the user with a password for SMB. We will have different ways of doing this with loops in the final review, which will allow you to vault passwords and lists of users. This is not recommended as everything is in plain text. This playbook is an example of one way to accomplish the change:

```
- name: Add SMB user
  user:
    name: "smbuser"
    shell: "/sbin/nologin"
    create_home: no
    system: yes

- name: Set SMB password for user
  command: smbpasswd -s -a smbuser
  args:
    stdin: "redhat\nredhat"
```

This portion of the code adds the group that we are using for SMB to the user:

```
- name: Add SMB users to the groups that are allowed
  user:
    name: "smbuser"
    group: "admins"
    append: "true"
```

Here we have created the directory and given the correct SELinux context:

```
- name: Create the directory to share
  file:
    path: "/shared/smbfolder"
    owner: "root"
    group: "admins"
    mode: "2775"
    state: "directory"
    setype: "samba_share_t"
```

Here, we have added the information for our share folder to the smb.conf file:

```
- name: Add the directory to the smb.conf file
  blockinfile:
    path: /etc/samba/smb.conf
    block: |
      [sambasharefolder]
                path = /shared/smbfolder
                writeable = yes
                valid users = sambauser, @admins
                write list = @admins
```

This portion of the code enables the SMB service as well as restarting it:

```
- name: Start or restart SMB for changes and enable it
  service:
    name: "smb"
    state: "restarted"
    enabled: "true"
```

This portion of the code opens up the necessary firewall rules:

```
- name: Add firewall rules to allow connectivity out
  firewalld:
    service: "samba"
    state: "enabled"
    immediate: "true"
    permanent: "true"
```

You can then run the playbook on the system using this command:

```
[emcleroy@rhel3 smb_playbook]$ ansible-playbook -i inventory
smb_playbook.yml -u emcleroy -k --ask-become-pass
```

This will ask you for your `ssh` password and your `become` password for the system. Please ensure that on the Ansible Control node (in our case, `rhel3`), `sshpass` is installed in order to avoid password failures. These can be handled through Ansible Vault as well or through other password credential systems that are out of the scope of this exam.

Next, we are going to show the playbook from the client side so that we can mount the new SMB share we created. We are going to create a new playbook as follows:

```
---
- hosts: sambaclient
  become: true
  become_method: sudo
  tasks:
    - name: Install Samba and Samba-Client
      package:
        name:
          - samba
          - samba-client
          - cifs-utils
        state: latest

    - name: Create a credential txt file
      copy:
        content: "username=smbuser\npassword=redhat\n"
        dest: /tmp/creds.txt
        owner: "root"
        group: "root"
        mode: "0600"
```

This creates a file with your credentials. You can also vault this file for encryption and use variables:

```
    - name: mount directory created
      file:
        path: "/mnt/shared"
        state: directory
```

```
        owner: "root"
        group: "root"
    - name: Mount the SMB share.
      shell: "mount -t cifs //192.168.1.198/smbfolder/mnt/
 shared -o credentials=/tmp/creds.txt"
```

Please note that the IP address of the server may be different for your lab setup.

This will mount the SMB share with the credentials you passed to the file. We will explore other ways of doing all of this in the final review with the full video in the CiA videos. The CiA videos will display the manual and automated code being run and will be placed on Packt's website.

Summary

This brings us to the end of this chapter and network file sharing. We have talked about many types of storage up until this point, from iSCSI to SMB. They all have one thing in common and that is networking. We will be moving on to introduce you to the networking aspect of how these things interconnect in our next chapter. We want to ensure that everyone is aware of how RHEL networking is handled, how it is different from traditional networking such as Cisco and others, and how it can help you. We will go over exam objectives and ensure you know how to connect these servers to pass the storage we talked about.

Part 2:
Red Hat Linux 8 – Configuring and Maintaining Networking with Automation

In this part, you will learn to set up and maintain Linux networking functions manually and automatically. This will meet the objectives set to pass the Red Hat EX358 exam.

This part contains the following chapters:

- *Chapter 3, Network Services with Automation – Introduction to Red Hat Linux Networking*
- *Chapter 4, Link Aggregation Creation – Creating Your Own Link and Mastering the Networking Domain*
- *Chapter 5, DNS, DHCP, and IP Addressing – Gaining Deeper Knowledge of Red Hat Linux Networking*

3

Network Services with Automation – Introduction to Red Hat Linux Networking

This chapter will introduce you to the exciting world of Red Hat Enterprise Linux networking. We will go over the intricacies of setting up the network interfaces to start. This will be done by using the `nmtui` command, which opens a visual tool as the main option. With this knowledge, you will begin to understand how your server connects to the outside world through an understanding of network interfaces.

Further chapters will delve deeper into other network applications that can be used to automatically set things, such as the **domain name system** (DNS), and gain an IP through **Dynamic Host Configuration Protocol** (DHCP). In this chapter, we are going to showcase how to set up both static and dynamic configurations and ensure they start when the server is turned on. This will ensure you always have network connectivity to the world when you need it.

In this chapter, we're going to cover the following main topics:

- The beginning of your journey into Linux networking
- Getting to know what the different terms mean and how they apply to what you are trying to achieve
- Creating your basic network profile and getting online so you can get applications up and running
- Automation of network services using Ansible

Technical requirements

The following items are necessary for following along with the chapter if you would like to gain hands-on knowledge of the systems.

Setting up GitHub access

Please refer to *Chapter 1, Block Storage – Learning How to Provision Block Storage on Red Hat Enterprise Linux*, for information about how to gain access to this book's GitHub repository. You will find the Ansible automation playbooks for this chapter at the following link: `https://github.com/PacktPublishing/Red-Hat-Certified-Specialist-in-Services-Management-and-Automation-EX358-Exam-Guide/tree/main/Chapter03`. Remember these are the suggested playbooks and are not the only way you can write them to make the playbooks work for you.

You can always change them by using `raw`, `shell`, or `cmd` to achieve the same results, but we are demonstrating the best way to accomplish our goals. Also, keep in mind that we are not using the FCQN that is needed in future versions of Ansible, as that will not be part on the exam as it is testing against Ansible 2.9.

Setting up your lab environment for networking

Please refer to the lab environment setup from the previous chapters if not already completed. You can also use this opportunity to create snapshots of your environment so that you can revert to them and do the hands-on exercises over and over again. This is outlined in *Chapter 2, Network File Storage – Expanding Your Knowledge of How to Share Data* .

The beginning of your journey into Linux networking!

There are fewer things more important than networking in today's world. You may not realize it, but you are always connected wherever you go. This can be from your home internet to your cell phone or the Wi-Fi at your local coffee shop. These all run on networks that interconnect our worlds. This is what allows you to get to Google.com or Amazon.com to search for or buy things. Without networking, none of this would be possible as you couldn't reach your destination. We will talk about networking in more depth over the next three chapters, but really, networking is used in all of this book – from SAN connectivity for storage to when you are connecting to a database.

Getting to know what the different terms mean and how they apply to what you are trying to achieve!

We are going to go over many different terms, starting out with abbreviations such as **ETH**, which stands for **Ethernet**. This is the physical connection used to plug your tower computer into a switch or router. This leads to what ETH uses to connect to the world, **Internet Protocol** (**IP**), or think of it as part of your home address to the world. And the other part of the address is your **media access control** (**MAC**) address, which is your unique hardware identifying number that is part of the physical ETH port. Think of IP as your city, state, and zip code, and your MAC as your street address. With the IP and MAC address put together, you are able to send and receive information as you would through the postal service only at over 1,000 times the speed. First up, we are going to create a basic internet connection manually and through automation.

Creating your basic network profile

Here we will focus on getting online so you can get applications up and running! First, we are going to configure our interface directly through network scripts. We want to determine the interface name we would like to change with the following command:

```
[emcleroy@rhel1 ~]$ ip a
```

This is fully illustrated in the output of the command in the following screenshot:

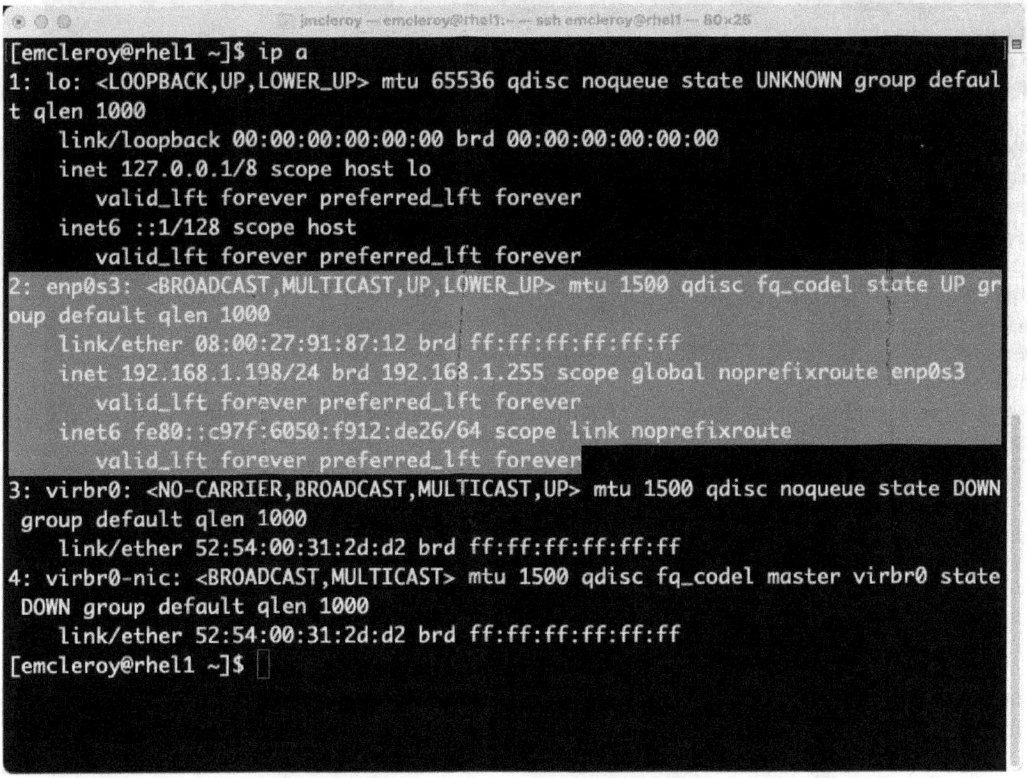

```
[emcleroy@rhel1 ~]$ ip a
1: lo: <LOOPBACK,UP,LOWER_UP> mtu 65536 qdisc noqueue state UNKNOWN group defaul
t qlen 1000
    link/loopback 00:00:00:00:00:00 brd 00:00:00:00:00:00
    inet 127.0.0.1/8 scope host lo
       valid_lft forever preferred_lft forever
    inet6 ::1/128 scope host
       valid_lft forever preferred_lft forever
2: enp0s3: <BROADCAST,MULTICAST,UP,LOWER_UP> mtu 1500 qdisc fq_codel state UP gr
oup default qlen 1000
    link/ether 08:00:27:91:87:12 brd ff:ff:ff:ff:ff:ff
    inet 192.168.1.198/24 brd 192.168.1.255 scope global noprefixroute enp0s3
       valid_lft forever preferred_lft forever
    inet6 fe80::c97f:6050:f912:de26/64 scope link noprefixroute
       valid_lft forever preferred_lft forever
3: virbr0: <NO-CARRIER,BROADCAST,MULTICAST,UP> mtu 1500 qdisc noqueue state DOWN
 group default qlen 1000
    link/ether 52:54:00:31:2d:d2 brd ff:ff:ff:ff:ff:ff
4: virbr0-nic: <BROADCAST,MULTICAST> mtu 1500 qdisc fq_codel master virbr0 state
 DOWN group default qlen 1000
    link/ether 52:54:00:31:2d:d2 brd ff:ff:ff:ff:ff:ff
[emcleroy@rhel1 ~]$
```

Figure 3.1 – Output from an IP command showing the interfaces

In our case, we will be manipulating the enp0s3 interface. This is the main interface, as the others are bridges and are not physical. The first interface is the loopback, which is used for the localhost connectivity for on-server communication.

Next, we can go to `/etc/sysconfig/network-scripts/`, where we will only find `ifcfg-enp0s3`. This is the interface configuration of the `enp0s3` connection. We can view it with the following command:

```
[emcleroy@rhel1 network-scripts]$ vi ifcfg-enp0s3
```

The output of the command looks like the following:

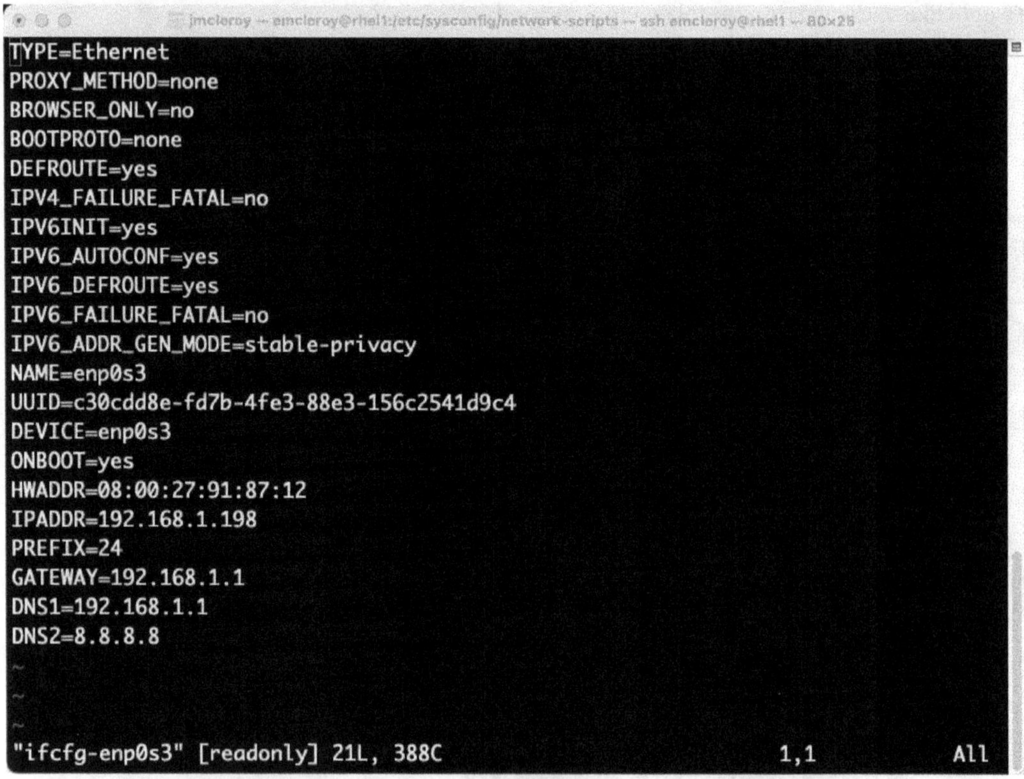

```
TYPE=Ethernet
PROXY_METHOD=none
BROWSER_ONLY=no
BOOTPROTO=none
DEFROUTE=yes
IPV4_FAILURE_FATAL=no
IPV6INIT=yes
IPV6_AUTOCONF=yes
IPV6_DEFROUTE=yes
IPV6_FAILURE_FATAL=no
IPV6_ADDR_GEN_MODE=stable-privacy
NAME=enp0s3
UUID=c30cdd8e-fd7b-4fe3-88e3-156c2541d9c4
DEVICE=enp0s3
ONBOOT=yes
HWADDR=08:00:27:91:87:12
IPADDR=192.168.1.198
PREFIX=24
GATEWAY=192.168.1.1
DNS1=192.168.1.1
DNS2=8.8.8.8
~
~
~
"ifcfg-enp0s3" [readonly] 21L, 388C                          1,1          All
```

Figure 3.2 – /etc/sysconfig/network-scripts/ifcfg-enp0s3 configuration

Please note that before you make any changes, you should make a copy of this file. Also, note that you will need to use the `sudo` command to escalate privileges have root privileges to configure this file:

```
[emcleroy@rhel1 network-scripts]$ sudo su
[root@rhel1 network-scripts]# cp ifcfg-enp0s3 ifcfg-enp0s3.old
```

After copying the file so that you have a backup, you can now make your changes to the interface configuration:

```
[root@rhel1 network-scripts]# vi ifcfg-enp0s3
```

An example of a change is to disable IPV6 DHCP, as shown in the following screenshot:

```
TYPE=Ethernet
PROXY_METHOD=none
BROWSER_ONLY=no
BOOTPROTO=none
DEFROUTE=yes
IPV4_FAILURE_FATAL=no
IPV6INIT=yes
IPV6_AUTOCONF=no
IPV6_DEFROUTE=yes
IPV6_FAILURE_FATAL=no
IPV6_ADDR_GEN_MODE=stable-privacy
NAME=enp0s3
UUID=c30cdd8e-fd7b-4fe3-88e3-156c2541d9c4
DEVICE=enp0s3
ONBOOT=yes
HWADDR=08:00:27:91:87:12
IPADDR=192.168.1.198
PREFIX=24
GATEWAY=192.168.1.1
DNS1=192.168.1.1
DNS2=8.8.8.8


-- INSERT --
```

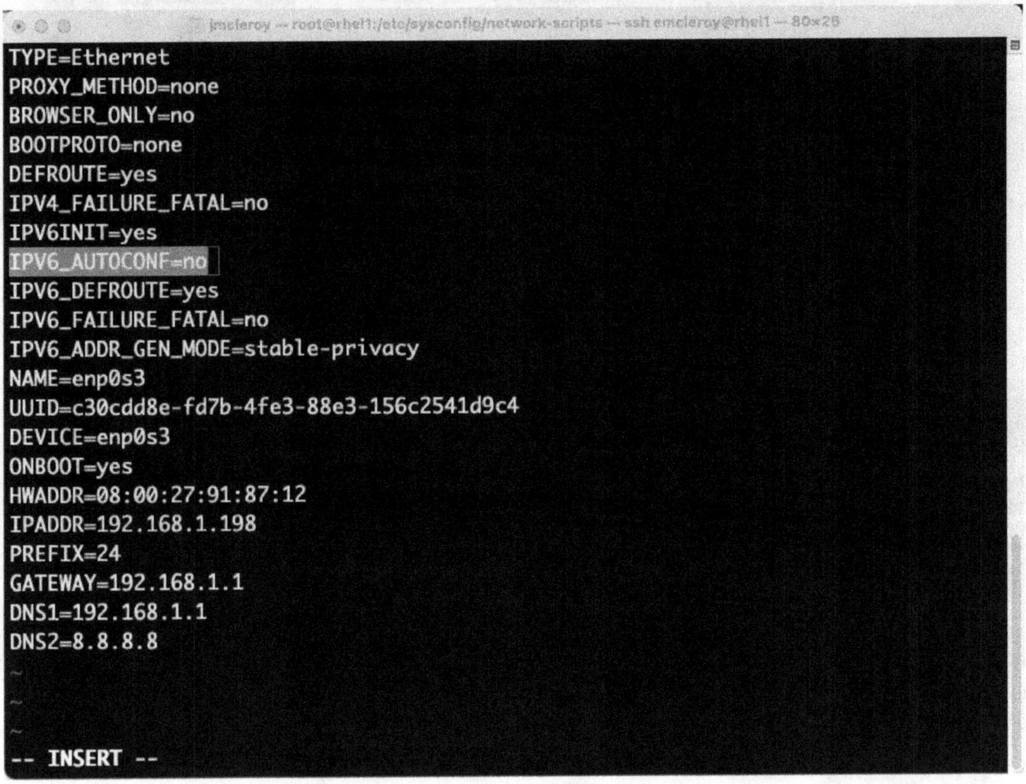

Figure 3.3 – A change made to the ifcfg-enp0s3 configuration file

This is just one of many things you can change. You can learn about all that you can modify using the supplied documentation during the test.

Secondly, we are going to use the **NetworkManager command line interface (nmcli)** to complete our setup of the interfaces. Through a series of commands, you can set up everything that can be found in /etc/sysconfig/network-scripts/ifcfg-enp0s3. Now, keep in mind there is a very well-documented **manual (man)** page for nmcli, as shown in the following screenshot:

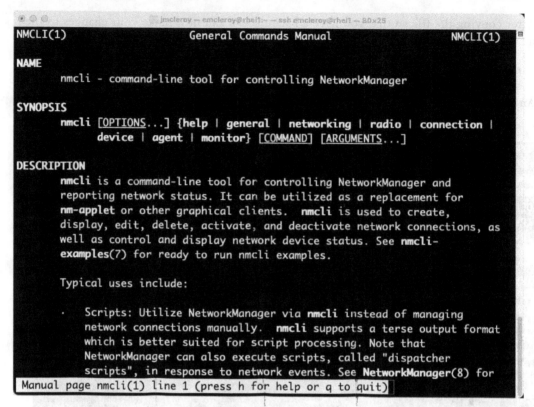

NMCLI(1)   General Commands Manual   NMCLI(1)

NAME
       nmcli - command-line tool for controlling NetworkManager

SYNOPSIS
       nmcli [OPTIONS...] {help | general | networking | radio | connection |
             device | agent | monitor} [COMMAND] [ARGUMENTS...]

DESCRIPTION
       nmcli is a command-line tool for controlling NetworkManager and
       reporting network status. It can be utilized as a replacement for
       nm-applet or other graphical clients. nmcli is used to create,
       display, edit, delete, activate, and deactivate network connections, as
       well as control and display network device status. See nmcli-
       examples(7) for ready to run nmcli examples.

       Typical uses include:

       ·   Scripts: Utilize NetworkManager via nmcli instead of managing
           network connections manually. nmcli supports a terse output format
           which is better suited for script processing. Note that
           NetworkManager can also execute scripts, called "dispatcher
           scripts", in response to network events. See NetworkManager(8) for
Manual page nmcli(1) line 1 (press h for help or q to quit)

Figure 3.4 – Man page for the nmcli command

This documentation will give you all of the information you need to control your interface. Let's try changing the enp0s3 DNS for our interface, as shown in the preceding screenshot. Using the man page, we can see what command we require in the examples to meet these variable command needs, as shown in the following screenshot:

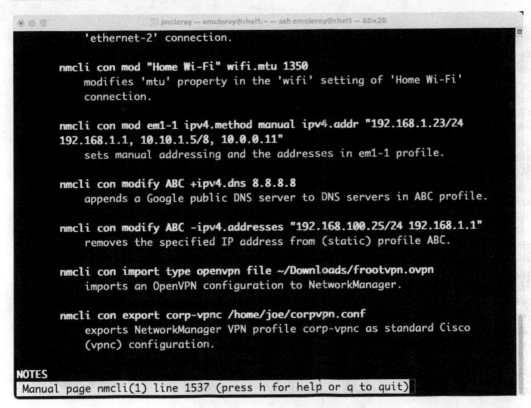

Figure 3.5 – Example of the required nmcli command string from the man page

As you can see, here we can use the following command to append a change to the dns:

```
[emcleroy@rhel1 ~]$ sudo nmcli con modify enp0s3 +ipv4.dns
4.4.4.4
```

This change has only taken effect in the startup script of the interface. This interface change can now be seen in the network scripts, as shown in the following screenshot:

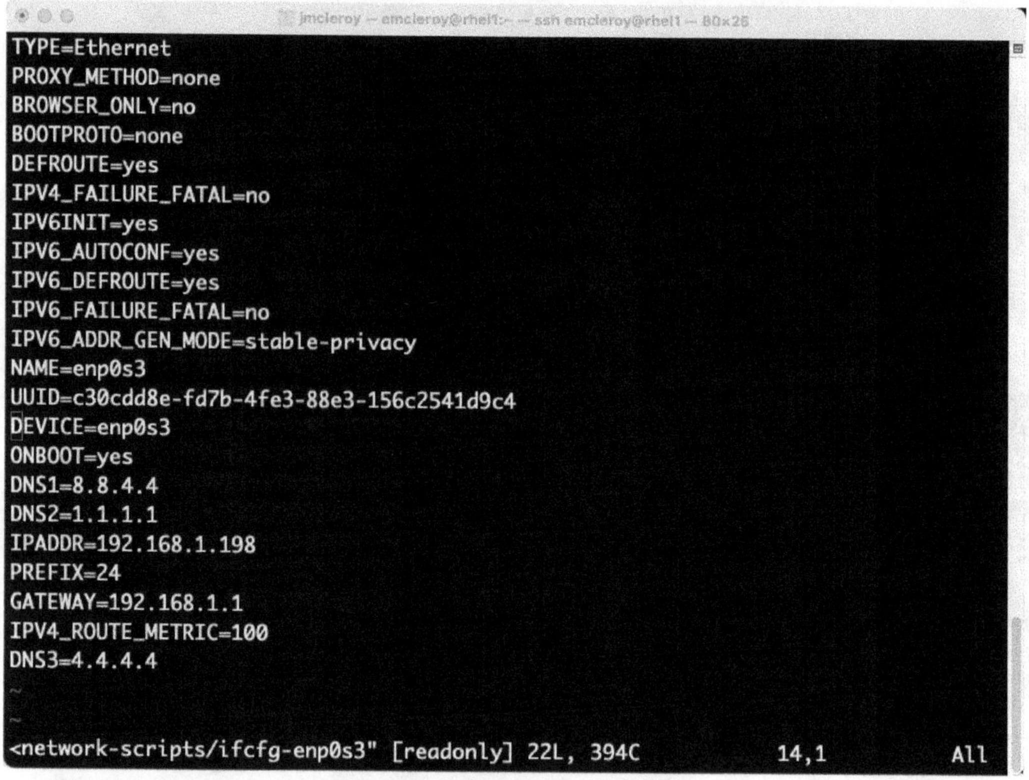

```
TYPE=Ethernet
PROXY_METHOD=none
BROWSER_ONLY=no
BOOTPROTO=none
DEFROUTE=yes
IPV4_FAILURE_FATAL=no
IPV6INIT=yes
IPV6_AUTOCONF=yes
IPV6_DEFROUTE=yes
IPV6_FAILURE_FATAL=no
IPV6_ADDR_GEN_MODE=stable-privacy
NAME=enp0s3
UUID=c30cdd8e-fd7b-4fe3-88e3-156c2541d9c4
DEVICE=enp0s3
ONBOOT=yes
DNS1=8.8.4.4
DNS2=1.1.1.1
IPADDR=192.168.1.198
PREFIX=24
GATEWAY=192.168.1.1
IPV4_ROUTE_METRIC=100
DNS3=4.4.4.4
~
~
~
<network-scripts/ifcfg-enp0s3" [readonly] 22L, 394C          14,1          All
```

Figure 3.6 – Results of the nmcli command string adding DNS3=4.4.4.4

If we want to make the interface change now, we need to bring down the interface and then bring it back up. This can be accomplished by using the following commands, but ensure that you are on the machine via **out-of-band management (OBM)**, as you will lose connectivity if this is your SSH connection to the server:

```
[emcleroy@rhel1 ~]# sudo nmcli connection down enp0s3
[emcleroy@rhel1 ~]# sudo nmcli connection up enp0s3
```

Next, we are going to use the graphical interface through the **NetworkManager text user interface (nmtui)** command:

```
[emcleroy@rhel1 ~]# nmtui
```

The result of running the nmtui command is shown in the following screenshot:

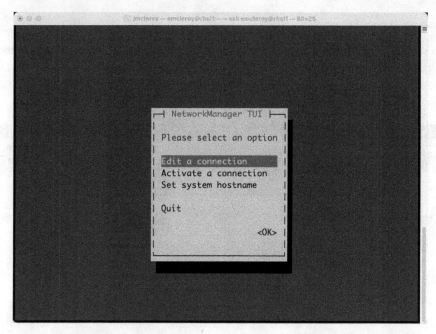

Figure 3.7 – The nmtui tool invoked, and options first presented

Choose **Edit a connection**:

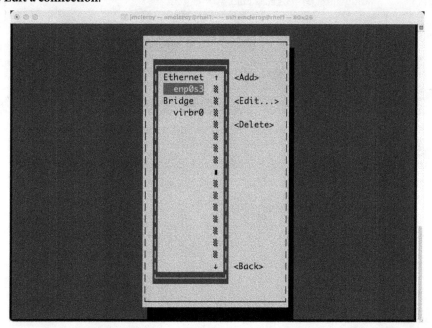

Figure 3.8 – Interfaces that are already present and available to change using nmtui

For this exercise, choose **Ethernet | enp0s3**.

As you can see here, **Profile name** matches the name of the interface. The **Device** field adds another element in the form of the MAC address. The **Automatic** configuration option dictates using a DHCP server to obtain your IP address automatically. We will get more into that in later chapters:

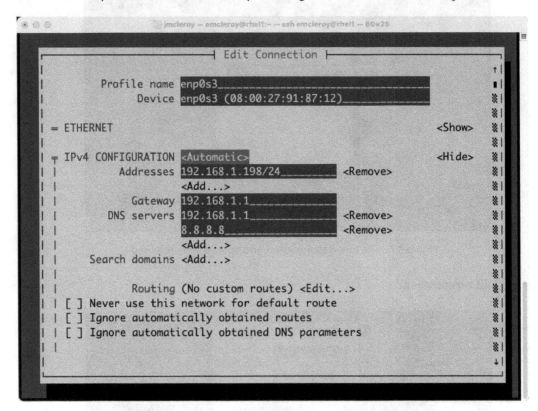

Figure 3.9 – Using nmtui to display the network interface and information

Here you can see the information, which in our case is manually set so that we do not have IP changes in order to ensure the following:

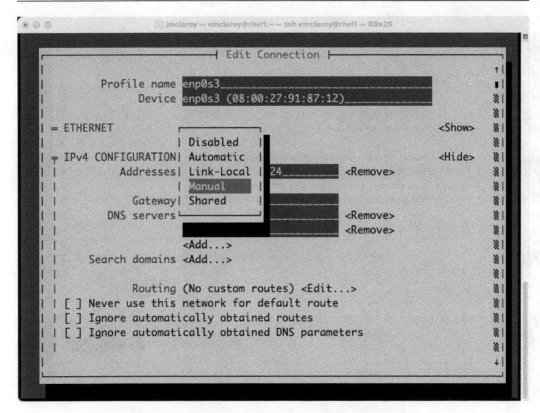

Figure 3.10 – Using nmtui to make a change

This is also where you set your gateway or the address for getting out to the rest of the network, in our case, 192.168.1.1, and DNS servers, and since our router sits on 192.168.1.1 also runs DNS from AT&T, for my purposes, it is my first DNS address. I follow this with a backup of a well-known DNS, such as Google's 8.8.8.8.

From there you can set other things, such as search domains and specific routing. These are a few more advanced areas that we don't need to cover right now.

Within this screen is where you can set your IPV6 information as well, such as automatically from a DHCP server or link-local to a manually set one:

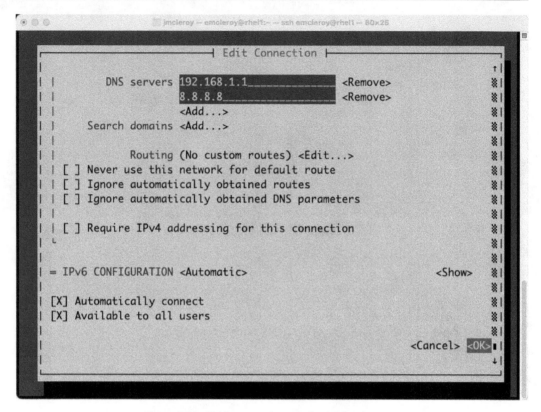

Figure 3.11 – Exiting nmtui correctly to save changes

All of the configuration settings look in order for our purposes, and any of them can be changed later if found to be incorrect. Just make sure that you hit the *Num Lock* key as needed, or it will not accept your input.

Once complete, choose **OK** or **Cancel**. We are going to cancel, as this is how we would like the interface setup to be.

Next, if you made changes you have to reset the connection for them to take effect. If you have direct access, you can go back to the **Activate a connection** option within the nmtui screen. This activation window is shown as follows in the nmtui GUI screenshot:

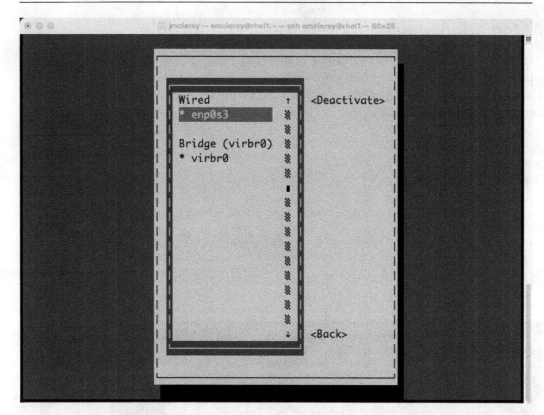

Figure 3.12 – Using nmtui to activate and deactivate network interfaces that have been created

To do this, choose **Deactivate** followed by **Reactivate**. You can only do this if you have direct access; if not, you will lose connectivity to your server. There is another way of testing this if you know your settings are correct or hope they are if you don't have direct server OBM. Just as with nmcli, we can repurpose one of those commands we learned earlier. This has to be run as root in order to ensure no password prompts or other things hang and cause the interface to remain down:

```
[emcleroy@rhel1 ~]$ sudo su -
[root@rhel1 ~]# nmcli networking off && nmcli networking on
```

Please understand that the recommendation is always to have remote onsite support or OBM, having had an unfortunate experience and a very long nervous car ride later.

Automation of network services using Ansible

Automating the setup of interfaces can be as simple or complex as you want to make it. Our recommendation is to start with installing the ansible galaxy system roles known as **Red Hat Enterprise Linux** (RHEL) system roles since we are using RHEL and not just any Linux. This can be accomplished by checking to see whether they are installed by ansible-galaxy first using the following command:

```
[emcleroy@rhel3 ~]$ ansible-galaxy list
# /usr/share/ansible/roles
# /etc/ansible/roles
```

This confirms that the network role is missing. Running the following command installs the roles:

```
[emcleroy@rhel3 ~]$ sudo dnf install rhel-system-roles
```

As you can see from this screenshot, the roles are now installed by running the $ ansible-galaxy list command:

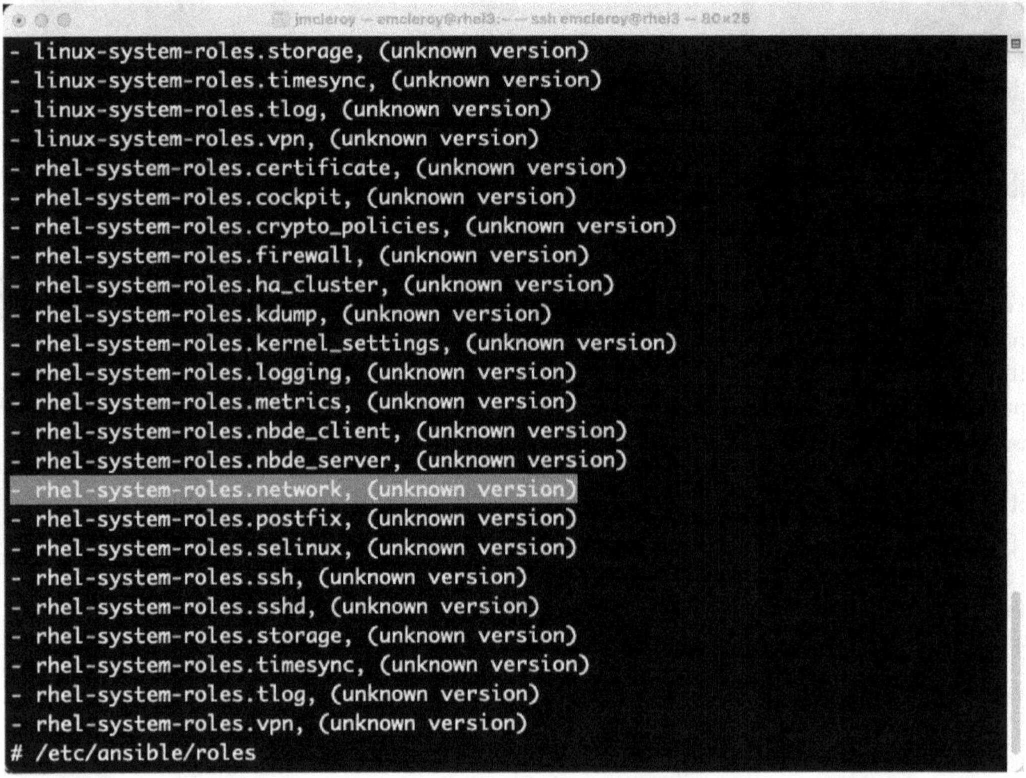

```
- linux-system-roles.storage, (unknown version)
- linux-system-roles.timesync, (unknown version)
- linux-system-roles.tlog, (unknown version)
- linux-system-roles.vpn, (unknown version)
- rhel-system-roles.certificate, (unknown version)
- rhel-system-roles.cockpit, (unknown version)
- rhel-system-roles.crypto_policies, (unknown version)
- rhel-system-roles.firewall, (unknown version)
- rhel-system-roles.ha_cluster, (unknown version)
- rhel-system-roles.kdump, (unknown version)
- rhel-system-roles.kernel_settings, (unknown version)
- rhel-system-roles.logging, (unknown version)
- rhel-system-roles.metrics, (unknown version)
- rhel-system-roles.nbde_client, (unknown version)
- rhel-system-roles.nbde_server, (unknown version)
- rhel-system-roles.network, (unknown version)
- rhel-system-roles.postfix, (unknown version)
- rhel-system-roles.selinux, (unknown version)
- rhel-system-roles.ssh, (unknown version)
- rhel-system-roles.sshd, (unknown version)
- rhel-system-roles.storage, (unknown version)
- rhel-system-roles.timesync, (unknown version)
- rhel-system-roles.tlog, (unknown version)
- rhel-system-roles.vpn, (unknown version)
# /etc/ansible/roles
```

Figure 3.13 – Screen seen after installing the system roles

You can get explanations on the different system roles or examples in the following location: `/usr/share/doc/rhel-system-roles-<version>/SUBSYSTEM/`. The command for network looks like the following:

```
[emcleroy@rhel3 network]$ cd /usr/share/doc/rhel-system-roles/
network/
```

You can then look at the README or any of the numerous examples. The README is illustrated as follows:

```
[emcleroy@rhel3 network]$ ls
example-bond_options-playbook.yml
example-bond_simple-playbook.yml
example-bond_with_vlan-playbook.yml
example-bridge_with_vlan-playbook.yml
example-down_profile+delete_interface-playbook.yml
example-down_profile-playbook.yml
example-dummy_simple-playbook.yml
example-eth_dns_support-playbook.yml
example-eth_simple_auto-playbook.yml
example-ethtool_coalesce-playbook.yml
example-ethtool_features_default-playbook.yml
example-ethtool_features-playbook.yml
example-ethtool_ring-playbook.yml
example-eth_with_802_1x-playbook.yml
example-eth_with_vlan-playbook.yml
example-infiniband-playbook.yml
example-inventory
example-ipv6_disabled-playbook.yml
example-macvlan-playbook.yml
example-match_path_support-playbook.yml
example-remove+down_profile-playbook.yml
example-remove_profile-playbook.yml
example-route_table_support-playbook.yml
example-team_simple-playbook.yml
```

Figure 3.14 – Examples of playbooks you can use to create the skeleton of your playbook

We can now set up an inventory for the playbook with the proper IP and Mac addresses. If they are not known, we can set different things. Let's start simple and just add the current IP of `rhel1` as shown in the following steps:

1. Move back to the home directory:

```
[emcleroy@rhel3 network]$ cd
```

2. Create a playbook folder:

```
[emcleroy@rhel3 ~]$ mkdir interface_playbook
```

3. Move into the working directory of the playbook folder:

```
[emcleroy@rhel3 ~]$ cd interface_playbook/
```

4. Create the inventory file:

```
[emcleroy@rhel3 interface_playbook]$ vi inventory
```

As shown in the following screenshot, we have the full inventory with a group for rhel1:

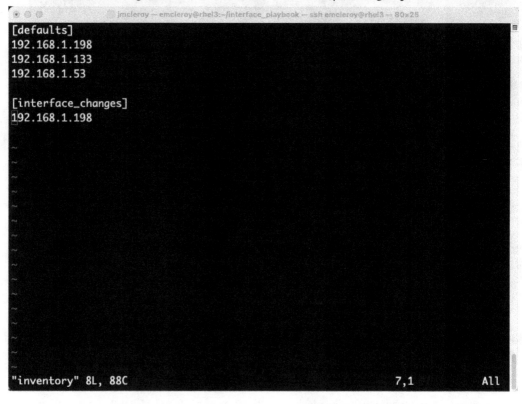

Figure 3.15 – Playbook inventory in order to pinpoint changes

Next, let's write a playbook based on one of the items in the `/usr/share/doc/rhel-system-roles/network/` directory. Let's make a change to the DNS record.

First, to see what a playbook looks like using the network roles since this will be our first introduction, please expand upon `/usr/share/doc/rhel-system-roles/network/example-eth_simple_auto-playbook.yml`. This is illustrated in the following screenshot:

```
[emcleroy@rhel3 interface_playbook]$ cat /usr/share/doc/rhel-system-roles/networ
k/example-eth_simple_auto-playbook.yml
# SPDX-License-Identifier: BSD-3-Clause
---
- hosts: network-test
  vars:
    network_connections:

      # Create one ethernet profile and activate it.
      # The profile uses automatic IP addressing
      # and is tied to the interface by MAC address.
      - name: prod1
        state: up
        type: ethernet
        autoconnect: yes
        mac: "{{ network_mac1 }}"
        mtu: 1450

  roles:
    - rhel-system-roles.network
[emcleroy@rhel3 interface_playbook]$ 
```

Figure 3.16 – Viewing an example in order to know what to do to set up the playbook

As you can see here, it is slightly different as we are laying out the variables and then the role runs the playbook. Take that structure and apply it to your playbook. You can see in the following screenshot that we have changed some of the items such as hosts and names to match our systems:

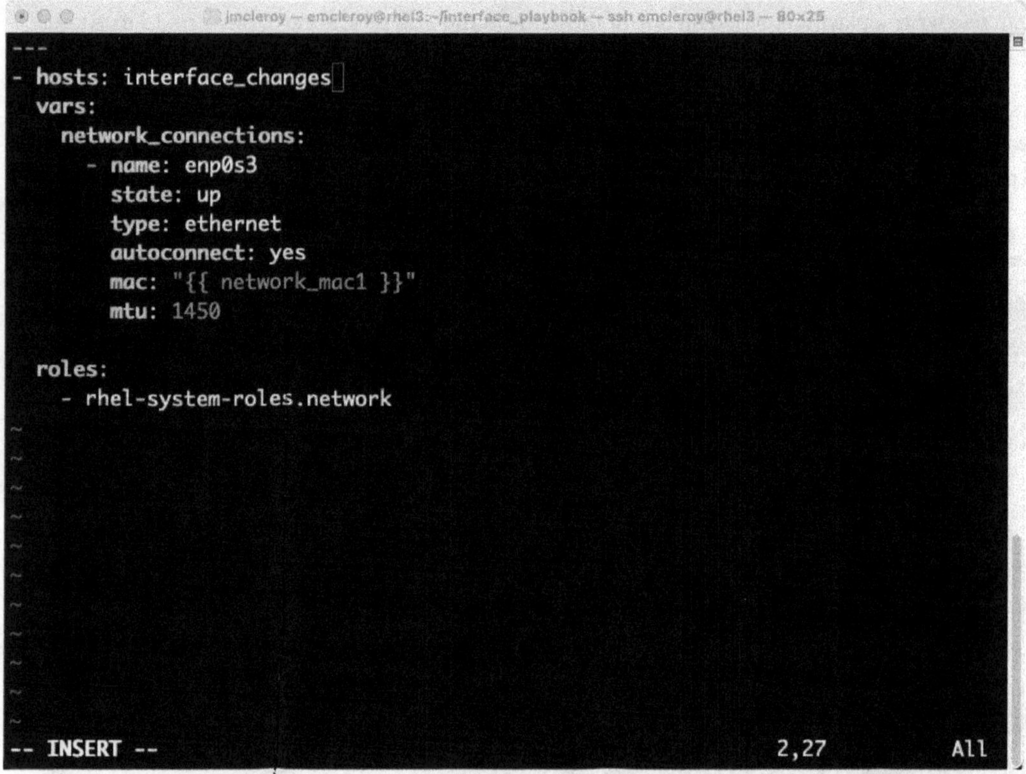

```
---
- hosts: interface_changes
  vars:
    network_connections:
      - name: enp0s3
        state: up
        type: ethernet
        autoconnect: yes
        mac: "{{ network_mac1 }}"
        mtu: 1450

  roles:
    - rhel-system-roles.network
~
~
~
~
~
~
~
~
~
~
-- INSERT --                                            2,27            All
```

Figure 3.17 – Using an example from /usr/share/doc/rhel-system-roles/network/ to start the playbook

After that, move on to checking out the README.md file for the variables you need. By exploring the README.md file, you can get examples towards the bottom, as shown in the following screenshot of a persistent connection:

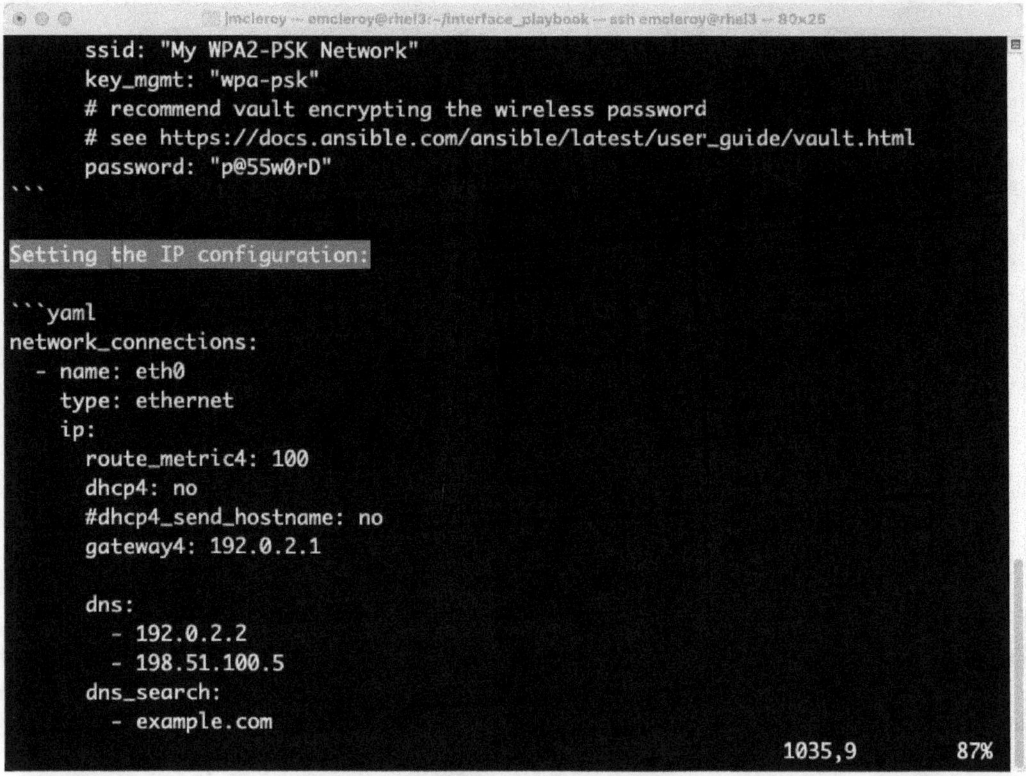

```
ssid: "My WPA2-PSK Network"
key_mgmt: "wpa-psk"
# recommend vault encrypting the wireless password
# see https://docs.ansible.com/ansible/latest/user_guide/vault.html
password: "p@55w0rD"
```


Setting the IP configuration:

```yaml
network_connections:
  - name: eth0
    type: ethernet
    ip:
      route_metric4: 100
      dhcp4: no
      #dhcp4_send_hostname: no
      gateway4: 192.0.2.1

      dns:
        - 192.0.2.2
        - 198.51.100.5
      dns_search:
        - example.com
```
                                                              1035,9   87%

Figure 3.18 – Using the README.md file from the systems role docs for more information

We are going to take that snippet and put it in our playbook, ensuring we change the information as needed (such as the MAC and IP addresses). We are going to combine the initial copy from the system and change it around to meet our static connection. This will allow us to add the DNS values. This can be seen as follows:

```
---
- hosts: interface_changes
  become: yes
  become_method: sudo
  vars:
    network_connections:
      - name: enp0s3
        state: up
        type: ethernet
        ip:
          route_metric4: 100
          dhcp4: no
          gateway4: 192.168.1.1

          dns:
            - 8.8.4.4
            - 1.1.1.1

          address:
            - 192.168.1.198/24

  roles:
    - rhel-system-roles.network

-- INSERT --                                              20,31            All
```

Figure 3.19 – Combining the two examples and the server information to create a complete playbook

Now, with all of the playbook options all set, we should be able to merge the changes to the interface, and it should have the new DNS records. Here are the records before running the playbook:

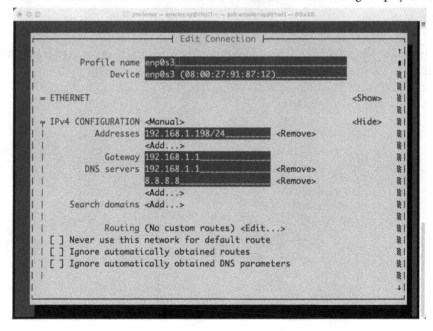

Figure 3.20 – Verifying that the changes are not in place

Next, we will run the playbook:

```
[emcleroy@rhel3 interface_playbook]$ ansible-playbook -i
inventory dns_change.yml -u emcleroy -k --ask-become-pass
```

As you can see in the following screenshot, it finished without any issues and shows one orange change, which should be our DNS entry:

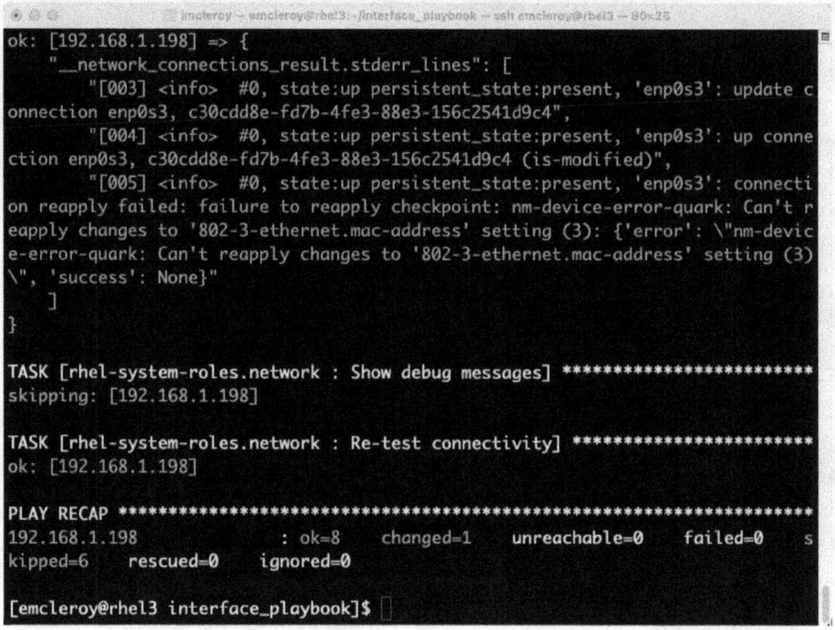

Figure 3.21 – Running the playbook

Now we can check the DNS records for `rhel1` and see whether they have changed correctly. As seen from the following screenshot, the changes were successful:

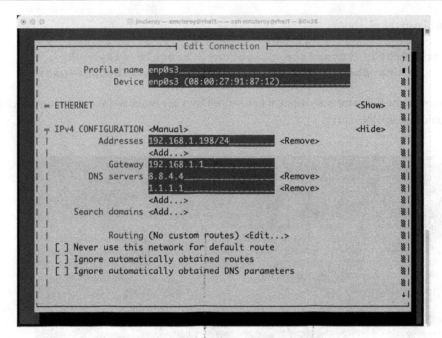

Figure 3.22 – Verifying the changes took effect

There are many changes you can use with the system roles that are not limited to just networking. These can be used for other parts of the exam and provide a helping hand that may be needed. So, keep in mind /usr/share/doc/rhel-system-roles/ from the installation we did previously, which is installed as a package.

Summary

This concludes our first steps into the world of networking. I hope learning about setting up your interfaces and changing how you talk to the world has helped you grow in some small way. In this chapter, you learned how to configure a network interface both manually and automatically. This allowed you to connect your server, thus, your application to the world using either a public or a private IP. Using tools such as nmtui and man pages make configuring these values more user-friendly. I hope we touched on all the parts you were particularly excited to see. We will go more in-depth into working with interfaces with **Link Aggregation** in our next chapter. This allows for many different traffic behaviors that can help you in your day-to-day job or if you need to set up failover, as I have in my home lab. For instance, if my primary internet connection goes down, my cellular network connection picks up where it left off. We are headed into more exciting areas of networking with many things to learn following that. Until the next chapter, breathe and remember this is a journey and it takes one step at a time. Also, don't forget that the documents are a part of the exam area. Just do not rely too heavily on them as time is key and is very limited.

Link Aggregation Creation – Creating Your Own Link and Mastering the Networking Domain

Network teaming is essential to **High Availability** (**HA**) and increasing throughput. HA is accomplished through normal teaming with nothing special needed from the switch. For more throughput, you need **Link Aggregation Control Protocol** (**LACP**) set up on the switch. Without the ability to have multiple network connectivity paths out of your systems to the internet, you are likely to get into an outage situation easily. Having multiple connectivity paths, such as with power being split between two providers, provides a better chance that you will not succumb to losing access to your application, thus alleviating some worry that the application that is presented to the world using what we learned earlier, in *Chapter 3, Network Services with Automation – Introduction to Red Hat Linux Networking*, will drop off the face of the planet and lose you money.

By utilizing network teaming in an HA function, if one connection goes down then the connectivity is not lost. When properly set up, such as having the two **Network Interface Cards** (**NICs**) going to different switches, it allows for fault tolerance. This is also a must when you are working on upgrades for the switches. If you don't have any redundancy, you will be down if the switch is going through maintenance. The same can be said about having your application on a single server, but we are here to talk about networks for now.

In this chapter, we're going to cover the following main topics:

- Getting to know link aggregation
- Creating different types of link aggregation profiles
- Taking the headache out of setting up your link aggregation by using Ansible Automation

Technical requirements

The technical requirements for this chapter are covered in the following two sections.

Setting up GitHub access

Please refer to *Chapter 1, Block Storage – Learning How to Provision Block Storage on Red Hat Enterprise Linux*, to gain access to GitHub instructions and you will find the Ansible Automation playbooks for this chapter at the following link: `https://github.com/PacktPublishing/Red-Hat-Certified-Specialist-in-Services-Management-and-Automation-EX358-Exam-Guide/tree/main/Chapter04`. Remember these are suggested playbooks and are not the only way you can write them to make the playbooks work for you.

You can always change them up using raw, shell, or cmd to achieve the same results but we are demonstrating the best way to accomplish the goals. Also keep in mind we are not using the FCQN that is needed in future versions of Ansible as that will not be supported in the exam as it tests against Ansible 2.9.

Setting up your lab environment for Network Interface Card (NIC) teaming

This has to be done for at least two of your systems so that you can test the different scenarios. I have opted to use Rhel1 and Rhel2 as Rhel3 is more like my workstation and Ansible control station. This will become apparent in the testing section and the exam tips section of the book. Find the instructions on setting up interfaces in VirtualBox as shown in the following screenshot.

Please ensure you turn off your virtual machines first, as follows:

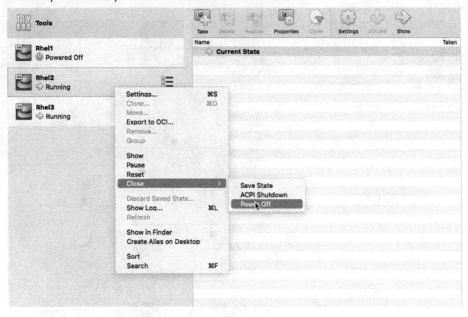

Figure 4.1 – Shutting down your VirtualBox VM

Then, choose **Settings**, as shown in the following screenshot:

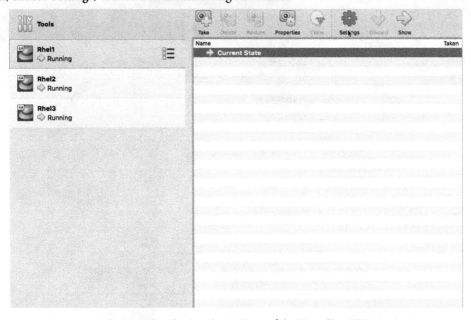

Figure 4.2 – Altering the settings of the VirtualBox VM

Then, we are going to choose **Network** and add adapter number 2, as follows:

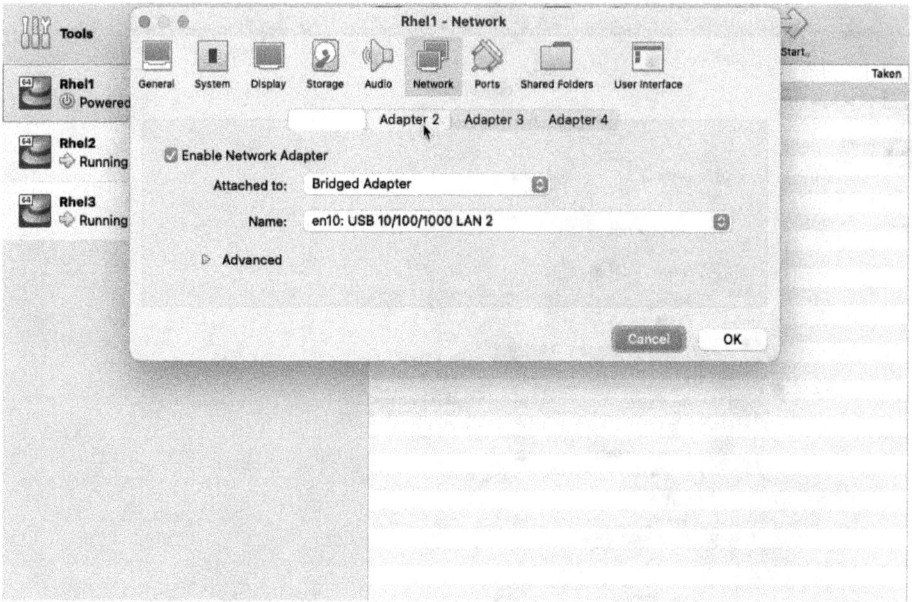

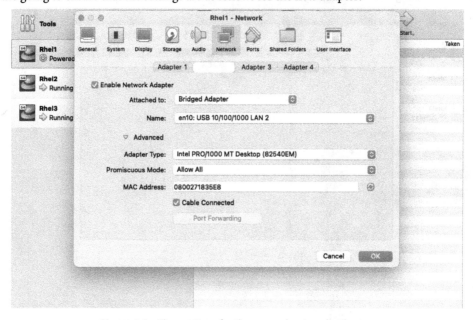

Figure 4.3 – Adding a second network adapter

We are going to ensure that the settings are as follows for the new adapter:

Figure 4.4 – The settings for the second network adapter

Finally, we are going to add a third adapter as follows:

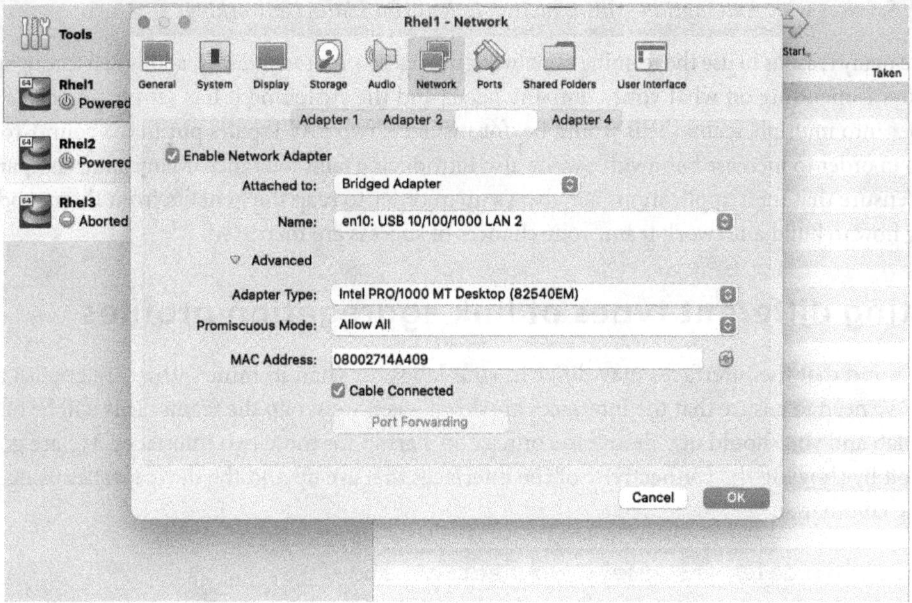

Figure 4.5 – The third and last network adapter added with the correct configuration

The reason for the two additional adapters is to maintain SSH connectivity if you are not using the normal start of the VM, which opens a GUI, but instead headless VM startup, which does not open a GUI in VirtualBox.

Getting to know link aggregation

Teamd handles the control of network teaming. It uses runners for this purpose. The list of teamd runners can be found here: https://access.redhat.com/documentation/en-us/ red_hat_enterprise_linux/8/html/configuring_and_managing_networking/ configuring-network-teaming_configuring-and-managing-networking. However, the ones we are going to concentrate on are as follows:

Runner Type	Description
lacp	The runner uses 802.3ad LACP.
loadbalance	The ports use a hash function to attempt to reach a balance between the ports.
activebackup	One port is set to active and the other is a backup for failover.
roundrobin	In this state, the ports are utilized in a round-robin state.

Table 4.1 – Some link aggregation profile types

For this task and to create network teams, we will be using the `nmcli` tool again, as we did in *Chapter 3, Network Services with Automation – Introduction to Red Hat Linux Networking*.

There are many reasons to use the teaming of network interfaces – for redundancy, additional throughput, and so on. Depending on what your company needs and the switching it has set up, you can even go deeper into multiple teams. This would be, for instance, two LACP pairs put into a round-robin scenario in order to increase bandwidth while also introducing fault tolerance. Many large companies need to ensure that their applications are always up in order to reap the benefits from them, and by learning how to build a network team, your chances of success are increased.

Creating different types of link aggregation profiles

Keep in mind that the interfaces may differ in your lab setup than in mine – they are enp0s8 and enp0s9. We need to ensure that the interfaces are down when we set up the teams. This will be based on your lab and you should not choose to connect on startup for these two interfaces. We are going to start off by showing the connectivity of the interfaces that are up and the device status using the following command:

```
[emcleroy@rhel1 ~]$ nmcli connection show
```

The output can be seen in the following screenshot:

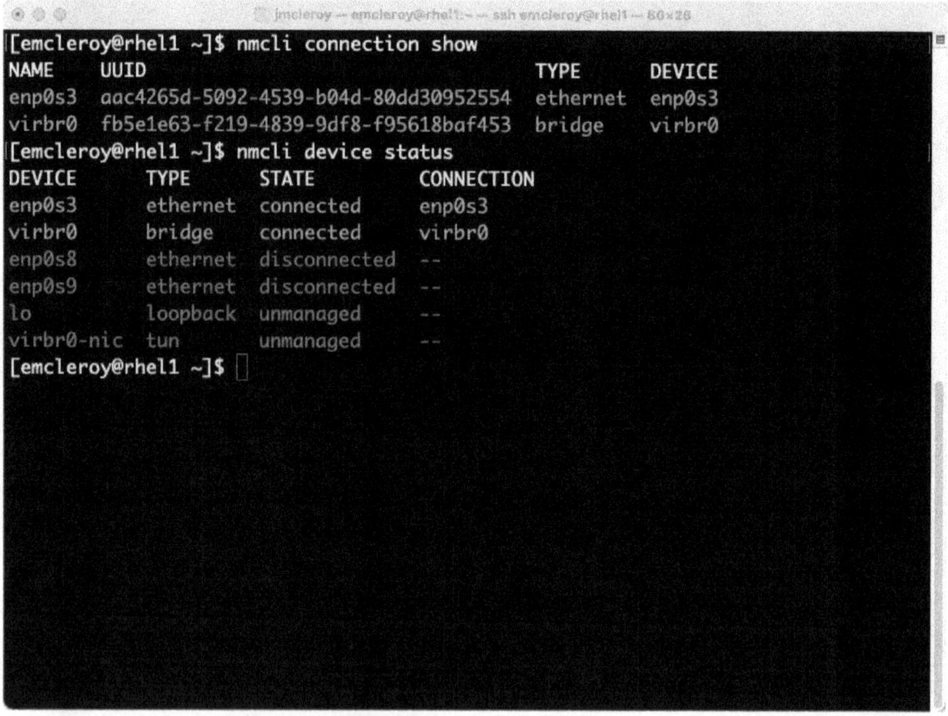

Figure 4.6 – Network interface connectivity, both virtual and physical

Next, we are going to start creating a team with the name bond1. We will use the following command:

```
[root@rhel1 emcleroy]# nmcli connection add type team con-name
bond1 ifname bond1 team.runner activebackup
```

This is illustrated in the following screenshot:

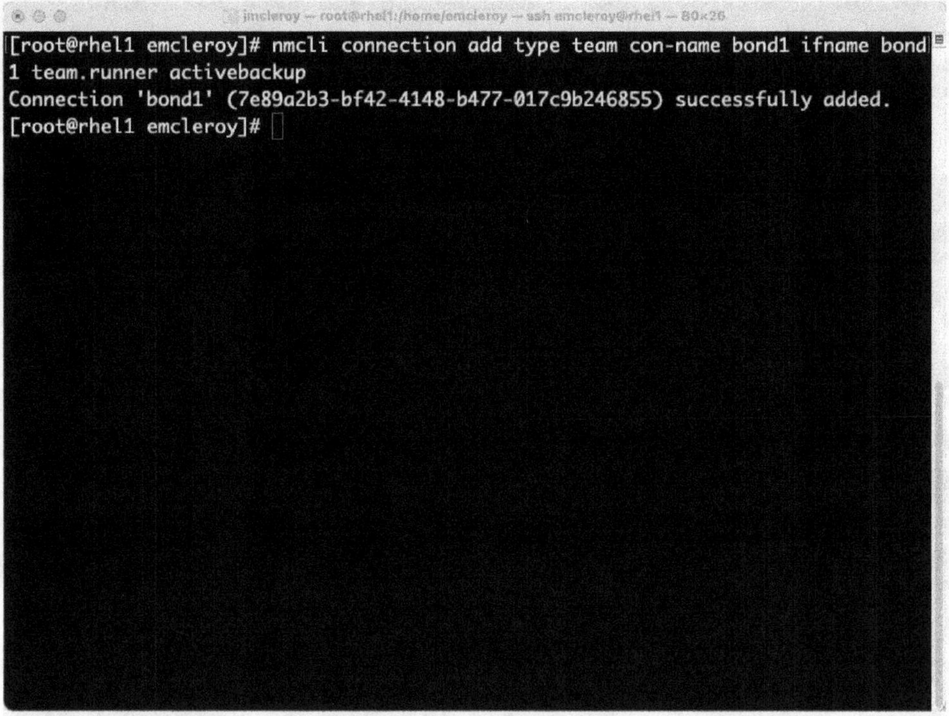

```
[root@rhel1 emcleroy]# nmcli connection add type team con-name bond1 ifname bond
1 team.runner activebackup
Connection 'bond1' (7e89a2b3-bf42-4148-b477-017c9b246855) successfully added.
[root@rhel1 emcleroy]#
```

Figure 4.7 – Full team command to create a team with name bond1 and runner of activebackup

Let's break down the screenshot we just looked at further. First, nmcli is the tool used to create the team. The code snippet of connection add is showing we are adding a new connection or interface type is showing that we are making a team in this case. The connection name (con-name) is the name you give it, which can be of your choosing. The interface name (ifname) – bond1, in this use case – is the name you give the interface. team.runner is the type of team you are creating; in our case, we are using activebackup, which means that one interface is passing traffic and the other is standing by in case of a failover. This can be due to manual intervention or if the interface loses connectivity.

The physical interfaces are then needed to bring the team up. We are using additional commands and modifying the team after it has been created. This allows us to add the interfaces as slaves to the team. We will use the following commands:

```
[root@rhel1 emcleroy]# nmcli connection add type team-slave
con-name bond1-enp0s8 ifname enp0s8 master bond1
[root@rhel1 emcleroy]# nmcli connection add type team-slave
con-name bond1-enp0s9 ifname enp0s9 master bond1
```

The following screenshot illustrates those commands:

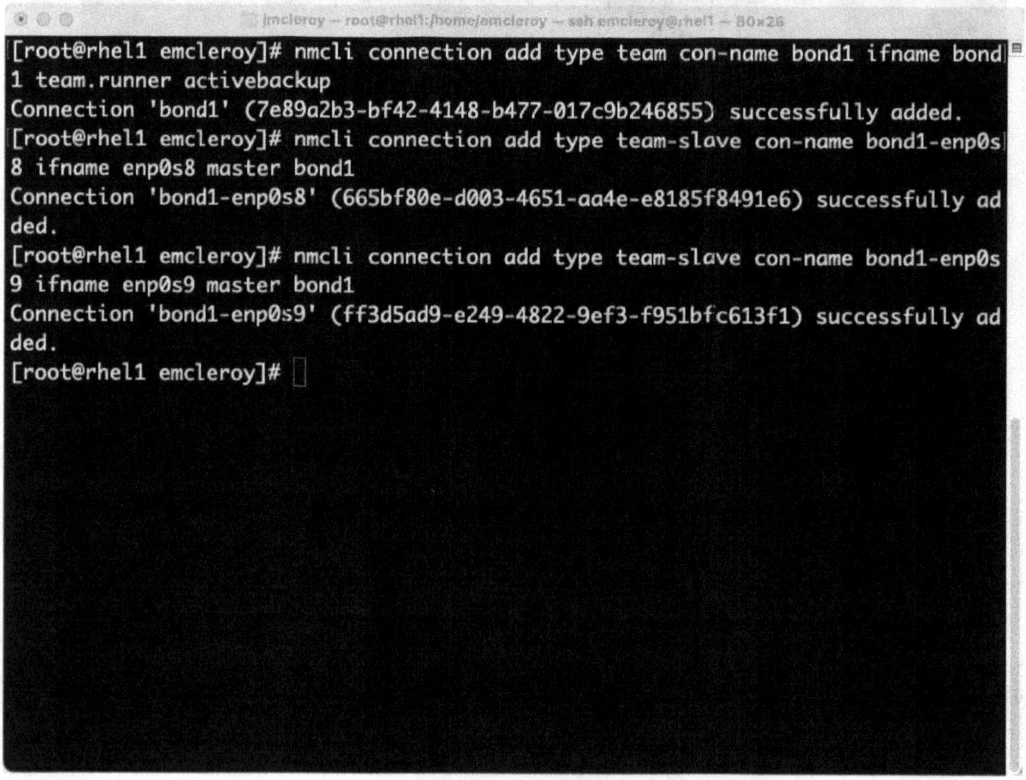

```
[root@rhel1 emcleroy]# nmcli connection add type team con-name bond1 ifname bond
1 team.runner activebackup
Connection 'bond1' (7e89a2b3-bf42-4148-b477-017c9b246855) successfully added.
[root@rhel1 emcleroy]# nmcli connection add type team-slave con-name bond1-enp0s
8 ifname enp0s8 master bond1
Connection 'bond1-enp0s8' (665bf80e-d003-4651-aa4e-e8185f8491e6) successfully ad
ded.
[root@rhel1 emcleroy]# nmcli connection add type team-slave con-name bond1-enp0s
9 ifname enp0s9 master bond1
Connection 'bond1-enp0s9' (ff3d5ad9-e249-4822-9ef3-f951bfc613f1) successfully ad
ded.
[root@rhel1 emcleroy]# 
```

Figure 4.8 – Adding physical interfaces to the virtual team

Additional items are needed to allow connectivity out to the world, as we learned in previous chapters. We will use the following commands to accomplish this change:

```
[root@rhel1 emcleroy]# nmcli connection modify bond1 ipv4.
addresses 192.168.1.233/24
[root@rhel1 emcleroy]# nmcli connection modify bond1 ipv4.
gateway 192.168.1.1
```

```
[root@rhel1 emcleroy]# nmcli connection modify bond1 ipv4.dns
192.168.1.1
[root@rhel1 emcleroy]# nmcli connection modify bond1 ipv4.
method manual
```

This is illustrated in the following screenshot:

```
jmcleroy — root@rhel1:/home/emcleroy — ssh emcleroy@rhel1 — 80×25
[root@rhel1 emcleroy]# nmcli connection modify bond1 ipv
ipv4.addresses              ipv4.may-fail            ipv6.dns-priority
ipv4.dad-timeout            ipv4.method              ipv6.dns-search
ipv4.dhcp-client-id         ipv4.never-default       ipv6.gateway
ipv4.dhcp-fqdn              ipv4.route-metric         ipv6.ignore-auto-dns
ipv4.dhcp-hostname          ipv4.routes              ipv6.ignore-auto-routes
ipv4.dhcp-send-hostname     ipv4.route-table         ipv6.ip6-privacy
ipv4.dhcp-timeout           ipv4.routing-rules       ipv6.may-fail
ipv4.dns                    ipv6.addresses           ipv6.method
ipv4.dns-options            ipv6.addr-gen-mode       ipv6.never-default
ipv4.dns-priority           ipv6.dhcp-duid           ipv6.route-metric
ipv4.dns-search             ipv6.dhcp-hostname       ipv6.routes
ipv4.gateway                ipv6.dhcp-send-hostname  ipv6.route-table
ipv4.ignore-auto-dns        ipv6.dns                 ipv6.routing-rules
ipv4.ignore-auto-routes     ipv6.dns-options         ipv6.token
[root@rhel1 emcleroy]# nmcli connection modify bond1 ipv4.addresses 192.168.1.23
3/24
[root@rhel1 emcleroy]# nmcli connection modify bond1 ipv4.gateway 192.168.1.1
[root@rhel1 emcleroy]# nmcli connection modify bond1 ipv4.dns 192.168.1.1
[root@rhel1 emcleroy]# nmcli connection modify bond1 ipv4.method manual
[root@rhel1 emcleroy]#
```

Figure 4.9 – Adding the additional configuration needed to the team

Finally, we want to ensure that after a reboot, the interface comes back up automatically so that you do not lose points in the exam. We will use this command to accomplish the task of ensuring the interface comes up after a reboot automatically:

```
[root@rhel1 emcleroy]# nmcli connection modify bond1
connection.autoconnect yes
```

This is shown in the following screenshot:

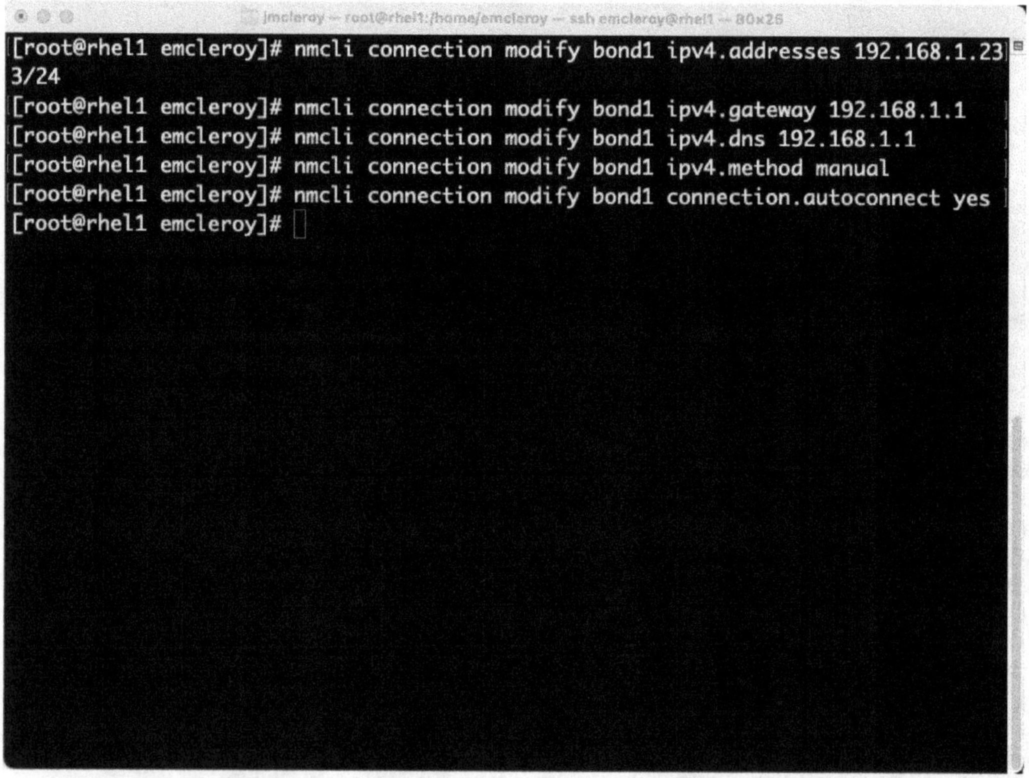

```
jmcleroy — root@rhel1:/home/emcleroy — ssh emcleroy@rhel1 — 80x25
[root@rhel1 emcleroy]# nmcli connection modify bond1 ipv4.addresses 192.168.1.23
3/24
[root@rhel1 emcleroy]# nmcli connection modify bond1 ipv4.gateway 192.168.1.1
[root@rhel1 emcleroy]# nmcli connection modify bond1 ipv4.dns 192.168.1.1
[root@rhel1 emcleroy]# nmcli connection modify bond1 ipv4.method manual
[root@rhel1 emcleroy]# nmcli connection modify bond1 connection.autoconnect yes
[root@rhel1 emcleroy]#
```

Figure 4.10 – Ensure that connectivity persists through reboots

You can run the following command to turn up the connection and ensure that your team is up and functioning correctly:

```
[root@rhel1 emcleroy]# nmcli con up bond1
[root@rhel1 emcleroy]# nmcli device status
```

In the following screen shot we can see the device status:

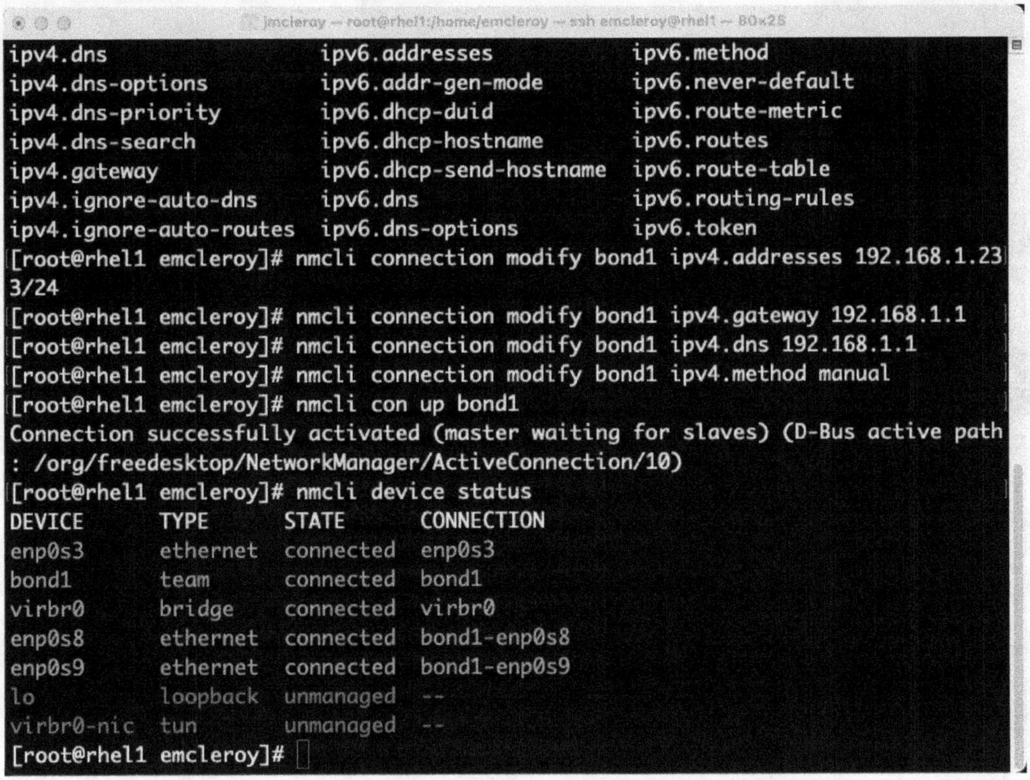
jmcleroy — root@rhel1:/home/emcleroy — ssh emcleroy@rhel1 — 80×25
ipv4.dns ipv6.addresses ipv6.method
ipv4.dns-options ipv6.addr-gen-mode ipv6.never-default
ipv4.dns-priority ipv6.dhcp-duid ipv6.route-metric
ipv4.dns-search ipv6.dhcp-hostname ipv6.routes
ipv4.gateway ipv6.dhcp-send-hostname ipv6.route-table
ipv4.ignore-auto-dns ipv6.dns ipv6.routing-rules
ipv4.ignore-auto-routes ipv6.dns-options ipv6.token
[root@rhel1 emcleroy]# nmcli connection modify bond1 ipv4.addresses 192.168.1.23
3/24
[root@rhel1 emcleroy]# nmcli connection modify bond1 ipv4.gateway 192.168.1.1
[root@rhel1 emcleroy]# nmcli connection modify bond1 ipv4.dns 192.168.1.1
[root@rhel1 emcleroy]# nmcli connection modify bond1 ipv4.method manual
[root@rhel1 emcleroy]# nmcli con up bond1
Connection successfully activated (master waiting for slaves) (D-Bus active path
: /org/freedesktop/NetworkManager/ActiveConnection/10)
[root@rhel1 emcleroy]# nmcli device status
DEVICE TYPE STATE CONNECTION
enp0s3 ethernet connected enp0s3
bond1 team connected bond1
virbr0 bridge connected virbr0
enp0s8 ethernet connected bond1-enp0s8
enp0s9 ethernet connected bond1-enp0s9
lo loopback unmanaged --
virbr0-nic tun unmanaged --
[root@rhel1 emcleroy]#

Figure 4.11 – Showing the device status of the team and team interfaces

The following command shows you what your team looks like, the connectivity, and which interface is currently active:

```
[root@rhel1 emcleroy]# teamdctl bond1 state
```

The output of that command is shown in the following screenshot:

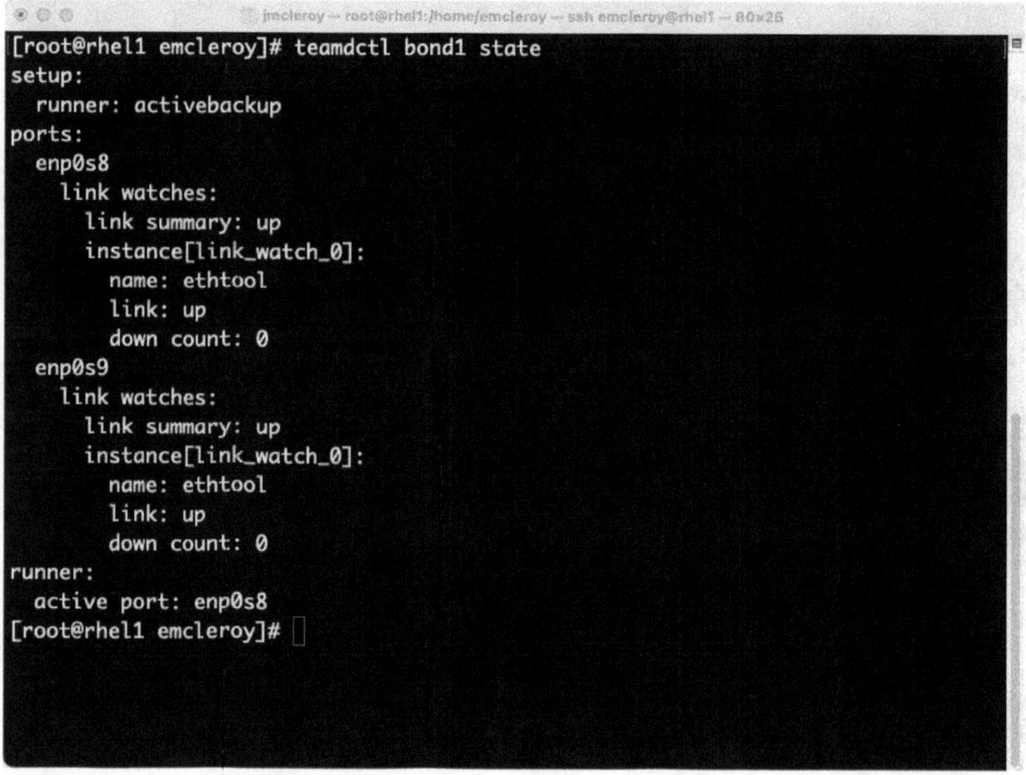

```
[root@rhel1 emcleroy]# teamdctl bond1 state
setup:
  runner: activebackup
ports:
  enp0s8
    link watches:
      link summary: up
      instance[link_watch_0]:
        name: ethtool
        link: up
        down count: 0
  enp0s9
    link watches:
      link summary: up
      instance[link_watch_0]:
        name: ethtool
        link: up
        down count: 0
runner:
  active port: enp0s8
[root@rhel1 emcleroy]#
```

Figure 4.12 – Team information and which port is currently active

Keep in mind if you are not able to access a server GUI, in most cases, you can install a GUI, but that will be discussed in later chapters. During your exam, you can use the nm-connection-editor command to launch a configurator. This can be seen as follows:

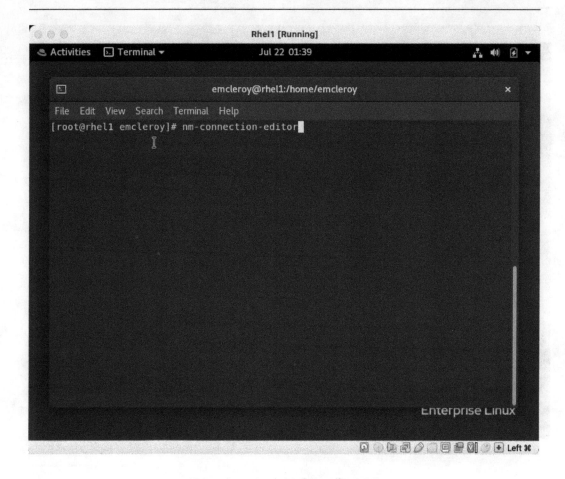

Figure 4.13 – Launching the command

After launching the command, you are then presented with a GUI that you can control. This allows you to choose a team and create all of the items you need for it without remembering the steps. First, choose a team, as follows:

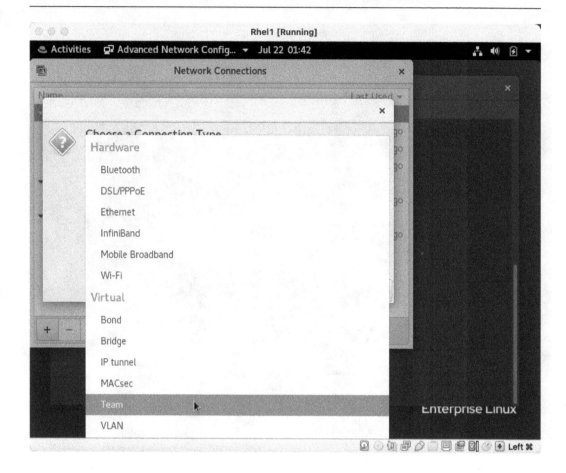

Figure 4.14 – Choose Team from the dropdown

Next, you will start to set the normal information for a team. As shown in the following screenshot, you are going to set the team name:

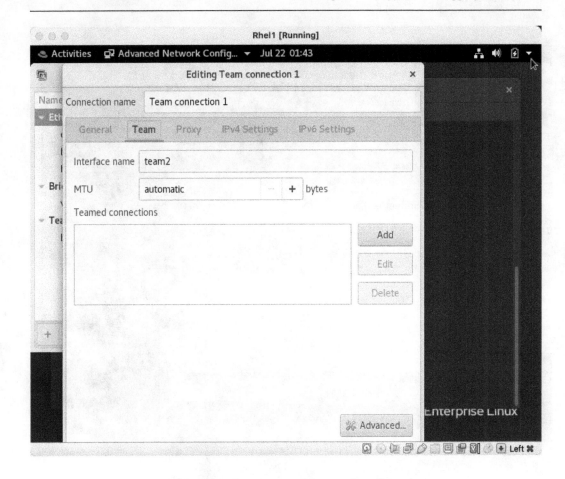

Figure 4.15 – Name the team and the interface of the team

After you set the team name, you are going to choose which physical interfaces or virtual interfaces will be a part of the new team. This is illustrated as follows:

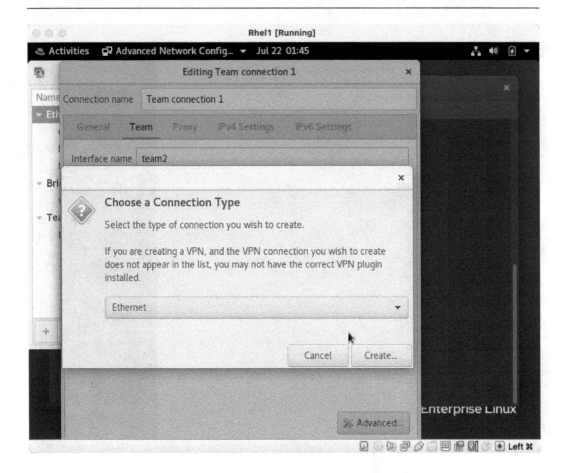

Figure 4.16 – Choose the interfaces to be a part of the team

You can even set different options for the interface you are adding to the team. For our purposes, we will not change anything and simply add it. This is illustrated in the following screenshot:

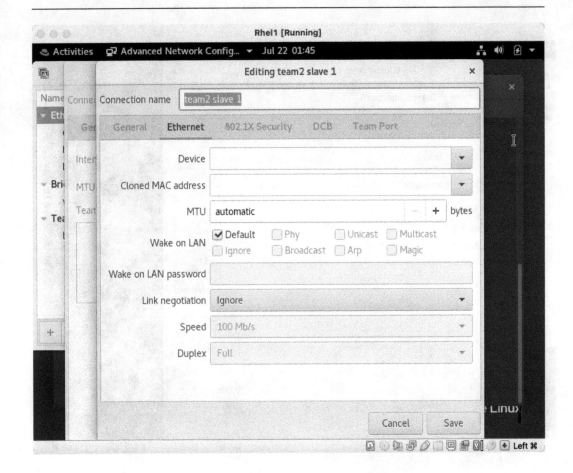

Figure 4.17 – The setting of interface settings at the slave interface level

You can see the outcome of adding both interfaces, or however many you would like to add to the team. This is shown as follows:

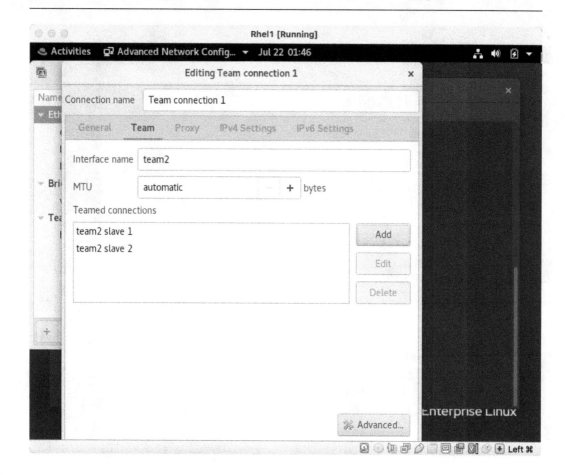

Figure 4.18 – Both interfaces after being added to the team

Finally, we can set the IP of the interface, the gateway, the DNS, and so on. This allows us to ensure that the team is now able to connect to the world. This is illustrated in the following screenshot:

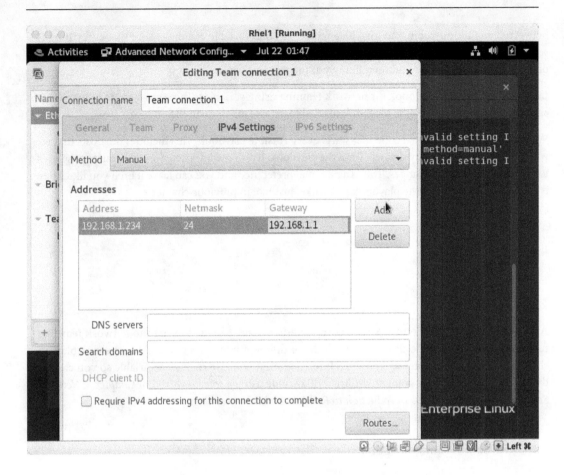

Figure 4.19 – Setting the information needed to talk to the world

That wraps up the manually configured way of doing things. Next up, we will make things easier by using Ansible Automation. Through Automation, we will be able to achieve the same results time and time again while lowering the possible human error factor. This will help when doing this type of task over and over again.

Taking the headache out of setting up your link aggregation by using Ansible Automation

Now, we all know having something automated allows us to save time and effort. It also allows us to be consistent and that is necessary, especially when you are working on networking. We can build multiple different playbooks to set up different things or we can use variables. In this case, we are going to use variables as they are easiest to change around. With a variable, you can change the different settings easily while still only needing one playbook. An example of that would be when you have to choose between using round-robin and load balancing. This one variable can make sure that everything is set correctly. Along with other variables, such as the IP address and gateways, you can get away with only writing one playbook and then filling out the rest with variables.

Let's start writing our playbook for network teaming using variables. These are a series of items that answer questions posed by the playbook needed to get the job done in the end.

We will then add our normal code that starts every Ansible playbook. This includes the hosts that we will be using from inventory that will be in the playbook. It also includes how we will escalate privileges to allow us to make the changes. This is a personal preference and you can leave it on if you like. Here is what the initial start of the playbook looks like, just like in previous chapters:

```
---

- name: Automated Teaming Setup
  hosts: rhel1
  become: true
  become_method: sudo
```

After we have the initial setup, we are going to add the next items to start the network teaming functions. This will include choosing a variable for different interfaces to use. Also, the variables IPs should use, along with gateway and DNS information. It will also include a variable so you can set the type of teaming we are going to do. This will include using a role, as we did in *Chapter 3, Network Services with Automation – Introduction to Red Hat Linux Networking*:

```
vars:
  network_connections:
    - name: bond1
      state: up
      type: team
      interface_name: bond1
  roles:
    - rhel-system-roles.network
```

You can see a man page about this role and how it works by going to `/usr/share/doc/rhel-system-roles/network/example-team_simple-playbook.yml` after you install the `rhel-system-roles.yml` file using `dnf install rhel-system-roles -y`. This shows you the variables you will need to fill out in the playbook to make the system role run correctly:

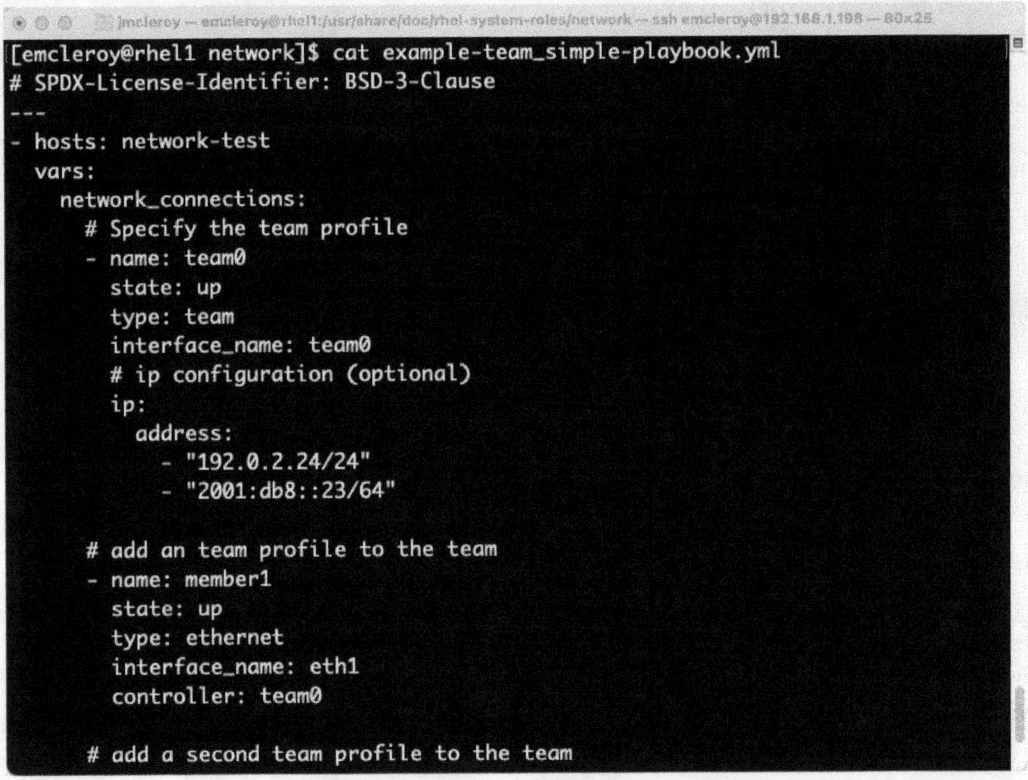

```
[emcleroy@rhel1 network]$ cat example-team_simple-playbook.yml
# SPDX-License-Identifier: BSD-3-Clause
---
- hosts: network-test
  vars:
    network_connections:
      # Specify the team profile
      - name: team0
        state: up
        type: team
        interface_name: team0
        # ip configuration (optional)
        ip:
          address:
            - "192.0.2.24/24"
            - "2001:db8::23/64"

      # add an team profile to the team
      - name: member1
        state: up
        type: ethernet
        interface_name: eth1
        controller: team0

      # add a second team profile to the team
```

Figure 4.20 – Rhel system role network teaming example

From here, we can use the information we gathered from the example to build the rest of our playbook. We will be inputting additional variables to be called by a variable file in the playbook directory instead of the raw values.

Here is the remaining code for the playbook to run successfully, and then we will see the variable file and hierarchy structure after the playbook code to ensure that the playbook can read the variables. This is shown in the following code snippet:

```
  vars:
    network_connections:
      - name: "{{ team_name }}"
        state: up
        type: team
        interface_name: "{{ team_name }}"
        ip:
          address:
```

```
              - "{{ ipv4_team_address }}"
              - "{{ ipv6_team_address }}"
      - name: member1
        state: up
        type: ethernet
        interface_name: "{{ int1 }}"
        controller: "{{ team_name }}"

      - name: member2
        state: up
        type: ethernet
        interface_name: "{{ int2 }}"
        controller: "{{ team_name }}"

  roles:
    - rhel-system-roles.network
```

This playbook will now create the build of the network team automatically. We will then create the variable file to answer the variables we created, such as `team_name`:

```
---
team_name: team1
ipv4_team_address: 192.168.1.233/24
ipv6_team_address:  fe80::9029:1a51:454:c2bd/64
int1: enp0s8
int2: enp0s9
```

This allows us to answer the variables but keep the playbook ambiguous to be used for multiple different uses instead of a one-and-done mentality. We will put this file in the directory with the playbook in the vars folder and name it `rhel1_team_vars.yml`. We will need to add this to the vars list in the playbook as well, so the playbook will know where to grab the information from:

```
  vars_files:
    - "{{ playbook_dir }}/vars/rhel1_team_vars.yml"

  vars:
```

```
network_connections:
  - name: "{{ team_name }}"
    state: up
    type: team
    interface_name: "{{ team_name }}"
    ip:
      address:
        - "{{ ipv4_team_address }}"
        - "{{ ipv6_team_address }}"

  - name: member1
    state: up
    type: ethernet
    interface_name: "{{ int1 }}"
    controller: "{{ team_name }}"

  - name: member2
    state: up
    type: ethernet
    interface_name: "{{ int2 }}"
    controller: "{{ team_name }}"

roles:
  - rhel-system-roles.network
```

Now, we can look at the layout of the playbook folder, which includes the inventory, vars file, and playbook:

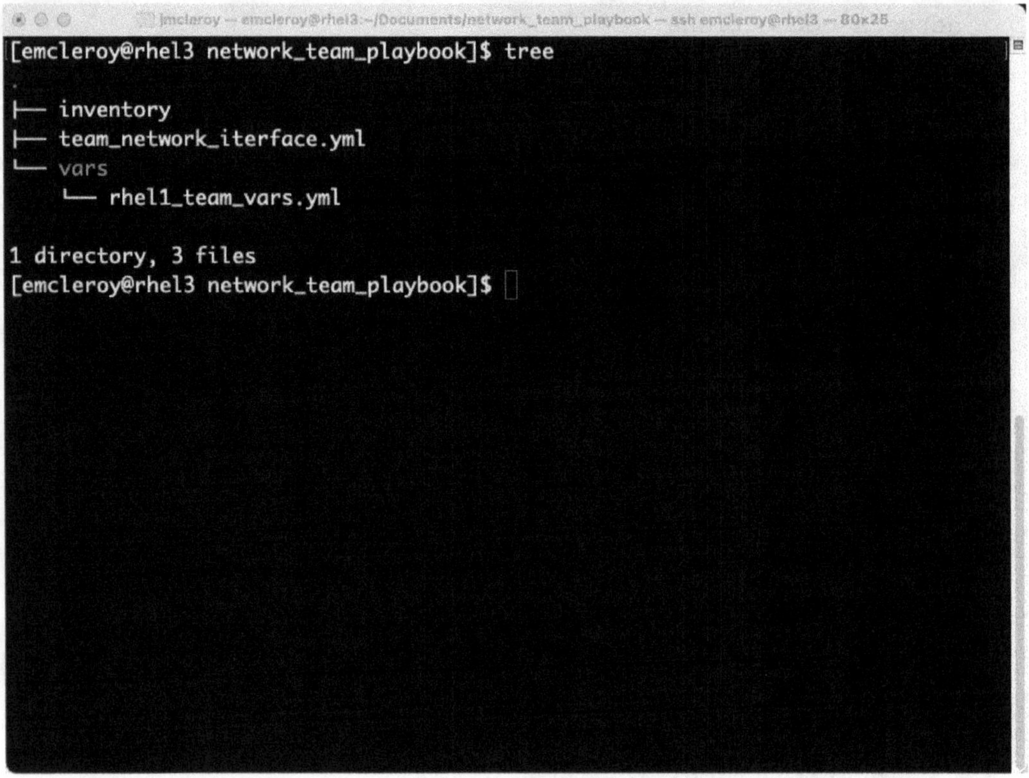

```
[emcleroy@rhel3 network_team_playbook]$ tree
.
├── inventory
├── team_network_iterface.yml
└── vars
    └── rhel1_team_vars.yml

1 directory, 3 files
[emcleroy@rhel3 network_team_playbook]$
```

Figure 4.21 – This shows the hierarchy of the playbook directory

For this playbook, the inventory will simply be `rhel1.example.com` under `defaults`, but it may differ in your lab setup. After that is set up, we will take a look at the interfaces before we set them up and then after the playbook has run.

Here, you can see that before we run the playbook, there are no teams in place:

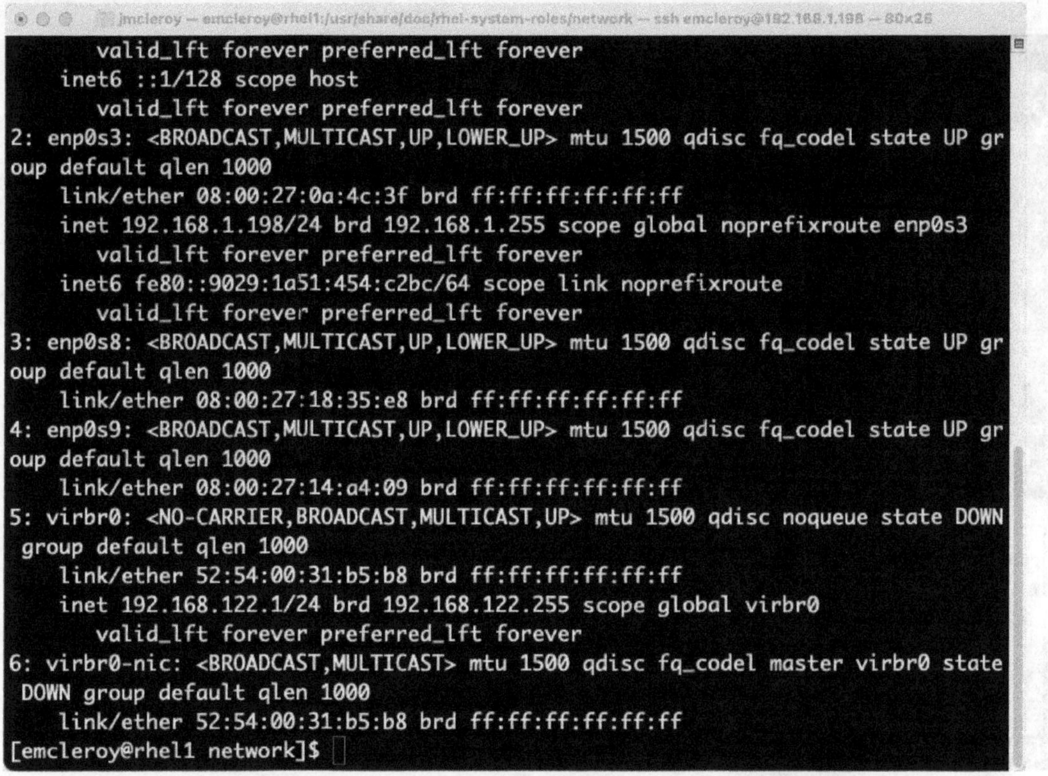

```
                jmcleroy — emcleroy@rhel1:/usr/share/doc/rhel-system-roles/network — ssh emcleroy@192.168.1.198 — 80x26
        valid_lft forever preferred_lft forever
    inet6 ::1/128 scope host
        valid_lft forever preferred_lft forever
2: enp0s3: <BROADCAST,MULTICAST,UP,LOWER_UP> mtu 1500 qdisc fq_codel state UP gr
oup default qlen 1000
    link/ether 08:00:27:0a:4c:3f brd ff:ff:ff:ff:ff:ff
    inet 192.168.1.198/24 brd 192.168.1.255 scope global noprefixroute enp0s3
        valid_lft forever preferred_lft forever
    inet6 fe80::9029:1a51:454:c2bc/64 scope link noprefixroute
        valid_lft forever preferred_lft forever
3: enp0s8: <BROADCAST,MULTICAST,UP,LOWER_UP> mtu 1500 qdisc fq_codel state UP gr
oup default qlen 1000
    link/ether 08:00:27:18:35:e8 brd ff:ff:ff:ff:ff:ff
4: enp0s9: <BROADCAST,MULTICAST,UP,LOWER_UP> mtu 1500 qdisc fq_codel state UP gr
oup default qlen 1000
    link/ether 08:00:27:14:a4:09 brd ff:ff:ff:ff:ff:ff
5: virbr0: <NO-CARRIER,BROADCAST,MULTICAST,UP> mtu 1500 qdisc noqueue state DOWN
 group default qlen 1000
    link/ether 52:54:00:31:b5:b8 brd ff:ff:ff:ff:ff:ff
    inet 192.168.122.1/24 brd 192.168.122.255 scope global virbr0
        valid_lft forever preferred_lft forever
6: virbr0-nic: <BROADCAST,MULTICAST> mtu 1500 qdisc fq_codel master virbr0 state
 DOWN group default qlen 1000
    link/ether 52:54:00:31:b5:b8 brd ff:ff:ff:ff:ff:ff
[emcleroy@rhel1 network]$
```

Figure 4.22 – The connections of the interfaces currently configured

Having seen the preceding connections, we are going to next run the playbook with the `ansible-playbook -i inventory team_network_interface.yml -u emcleroy -k --ask-become -v` command and view the team that was created:

```
    inet6 fe80::9029:1a51:454:c2bc/64 scope link noprefixroute
        valid_lft forever preferred_lft forever
3: enp0s8: <BROADCAST,MULTICAST,UP,LOWER_UP> mtu 1500 qdisc fq_codel master team
1 state UP group default qlen 1000
    link/ether 08:00:27:18:35:e8 brd ff:ff:ff:ff:ff:ff
4: enp0s9: <BROADCAST,MULTICAST,UP,LOWER_UP> mtu 1500 qdisc fq_codel master team
1 state UP group default qlen 1000
    link/ether 08:00:27:18:35:e8 brd ff:ff:ff:ff:ff:ff
5: virbr0: <NO-CARRIER,BROADCAST,MULTICAST,UP> mtu 1500 qdisc noqueue state DOWN
 group default qlen 1000
    link/ether 52:54:00:31:b5:b8 brd ff:ff:ff:ff:ff:ff
    inet 192.168.122.1/24 brd 192.168.122.255 scope global virbr0
        valid_lft forever preferred_lft forever
6: virbr0-nic: <BROADCAST,MULTICAST> mtu 1500 qdisc fq_codel master virbr0 state
DOWN group default qlen 1000
    link/ether 52:54:00:31:b5:b8 brd ff:ff:ff:ff:ff:ff
7: team1: <BROADCAST,MULTICAST,UP,LOWER_UP> mtu 1500 qdisc noqueue state UP grou
p default qlen 1000
    link/ether 08:00:27:18:35:e8 brd ff:ff:ff:ff:ff:ff
    inet 192.168.1.162/24 brd 192.168.1.255 scope global dynamic noprefixroute t
eam1
        valid_lft 86159sec preferred_lft 86159sec
    inet6 fe80::1f5f:2bb6:2bff:a23c/64 scope link noprefixroute
        valid_lft forever preferred_lft forever
[emcleroy@rhel1 ~]$
```

Figure 4.22 – Here, you can see the team was created

As you have learned in this section, Ansible Automation can make setting up a network team a breeze. Along with the ability to save time for other more important projects, Ansible Automation lowers the human error factor. This allows users to be confident that the configuration on each server will match. This allows faster troubleshooting and ease of deployment.

Summary

In this chapter, we have learned about network teaming. We covered how it can be used for HA and for throughput advantages. We worked on ensuring that you understand the value behind using a network team and the importance it has in a production environment. When you have multiple paths, for instance, you have the ability to take down a switch for maintenance without causing an issue to your running applications. These are just some of the advantages of network teaming and there are many more. In the next chapter, we will be digging further into the networking space for Rhel 8, in which we will discuss DNS, DHCP, and IPs in further detail. We will learn how to set them up and enable servers to provide functions such as DNS and DHCP. Take a breather, and then let's get ready to leap into more networking fun.

## 5

DNS, DHCP, and IP Addressing – Gaining Deeper Knowledge of Red Hat Linux Networking

In this chapter, we are going to dig deeper into networking. We will discuss **Internet Protocol (IP)** addressing in two ways: static and **Dynamic Host Configuration Protocol (DHCP)**. Static addressing occurs when you set the IP address, such as 192.168.1.10, with a gateway address of 192.168.1.1, whereas with DHCP, the IP information is provided by a DHCP server and the system goes through a handshake mechanism to obtain an IP and the DNS and gateway information on the connected subnet. We will go into more detail about static addressing and DHCP addressing later in this chapter. We will also be setting up a DHCP server to provide IP addressing information to systems on your network both manually and through Ansible Automation.

We are then going to look at how we can find other addresses on the internet through the **Dynamic Name System (DNS)**, which attaches **Fully Qualified Domain Names (FQDNs)** to IP addresses. This translates things such as http://redhat.com to an IP address that your computer can then use to reach from your web browser or other connectivity sources such as Telnet or SSH. This is all handled through domain name registrars where the DNS is associated with the website to IP (and vice versa) information. We will set up a DNS server both manually and through Ansible Automation.

In this chapter, we're going to cover the following main topics:

- Diving deeper into Linux networking where we look at DNS, DHCP, and static IP addressing
- Setting up static IP addresses for times when DHCP is not available but you still need to make that service reachable
- Using the basic out-of-the-box DHCP configuration to get online fast when available on your network
- Learning what the DNS is and why you need to know about it

The technical requirements for this chapter are covered in the following section.

Setting up GitHub access

Please refer to the instructions found in *Chapter 1, Block Storage – Learning How to Provision Block Storage on Red Hat Enterprise Linux*, to gain access to GitHub. You will find the Ansible Automation playbooks for this chapter at the following link: `https://github.com/PacktPublishing/Red-Hat-Certified-Specialist-in-Services-Management-and-Automation-EX358-Exam-Guide/tree/main/Chapter05`. Remember, these are suggested playbooks and are not the only way you can write them to make the playbooks work for you.

You can always change them up using raw, shell, or CMD to achieve the same results, but we are demonstrating the best way to accomplish our goals. Also keep in mind that we are not using the FQCN that is needed in the future version of Ansible, as that will not be supported in the exam since it is testing against Ansible 2.9.

Diving deeper into Linux networking where we look at DNS, DHCP, and static IP addressing

So far, we have briefly talked about Linux networking and what it means to you – the ability to provide and gain access to your servers, applications, and so on that reside on your and other networks. We will take that a step past the initial setup of the IP addressing of an interface and build upon what we have learned so far in this book. We will show you not only how your devices get your IP, gateway, and DNS configurations but also how to provide those services to your servers and applications.

First, we will go over the importance of each aspect of the different items. Your IP address is like your home address. Your gateway is like your driveway out into the world. The DNS is like your GPS because it shows you how to reach your online destination. The configuration of all of these things, either automatically through DHCP or manually through static addressing, is the bare minimum needed to get online and reach your destination.

We have talked previously about IP addressing and how static addressing is done manually through interface configuration. The same is true for DHCP; however, your system talks to another device to automatically obtain the information it needs to get online. This information includes the DNS, which is the GPS of how you locate the different things you are looking to find online, such as e-commerce websites, news, and so on. We will delve deeper into each of these categories as we move through this chapter.

Setting up static IP addresses for times when DHCP is not available but you still need to make that service reachable

This is for times when you don't have DHCP or when you need to always ensure the same address. This will provide the IP address, gateway, DNS, and routing that comes along with that setup; you sometimes need to enable the system via static configurations. You will notice that in the following screenshot, it did get a DHCP address but for our purposes, we are going to assume it received the incorrect address that is needed for the application. We will show the current IP address on the server using the following command:

```
[emcleroy@rhel2 ~]$ ip a | grep enp0s3
```

The following screenshot shows the IP address information that is currently configured on the server:

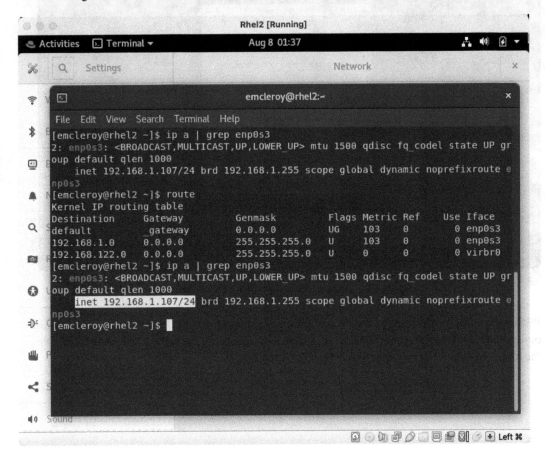

Figure 5.1 – IP addressing based on DHCP currently

So, in order to resolve this issue, we are going to set the system up statically using the nmtui command to gain a GUI-style interface, as shown in the following screenshot:

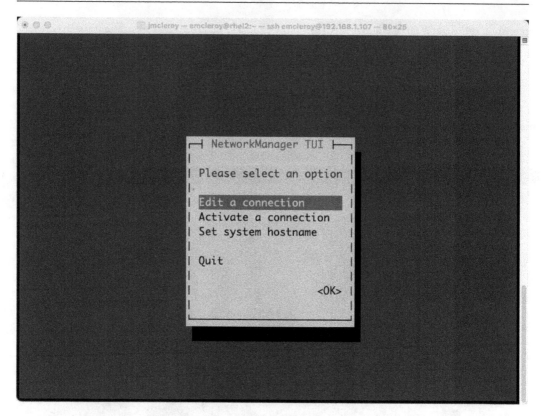

Figure 5.2 – Initiated GUI using the nmtui command

Within the GUI, we are going to choose the **Edit a connection** option. This will allow us to choose our interface and set the appropriate settings manually. Choosing the right interface after selecting the mentioned option is shown in the following screenshot:

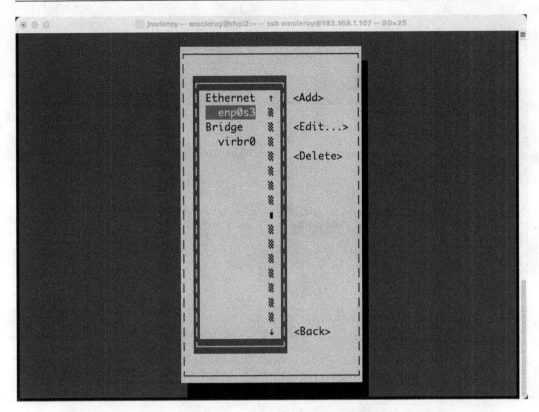

Figure 5.3 – The Edit a connection option chosen and the correct interface highlighted

After choosing the correct interface, we are going to choose the **Manual** setting so we can open up the needed fields for manipulation, as shown in the following screenshot:

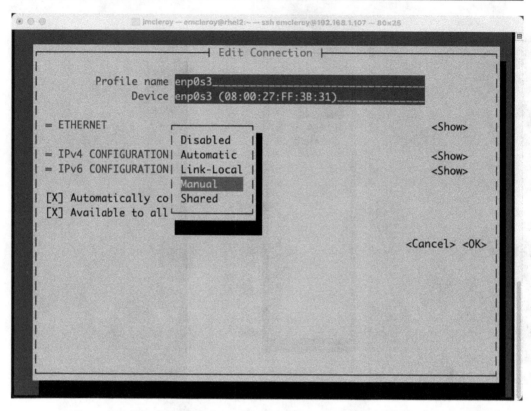

Figure 5.4 – Selecting Manual to open up the Ethernet settings for configuration

We will then set the correct configuration for the interface. This might be information that you have to obtain from your network team in some cases. The reason behind that is that you have to know your DNS, gateway, and subnet range. These may not be known to you as a server admin, and that's why DHCP is preferred, at least for desktop usage, as it auto-populates all of these fields. The following screenshot shows the required configuration to get this interface up and connected to the world:

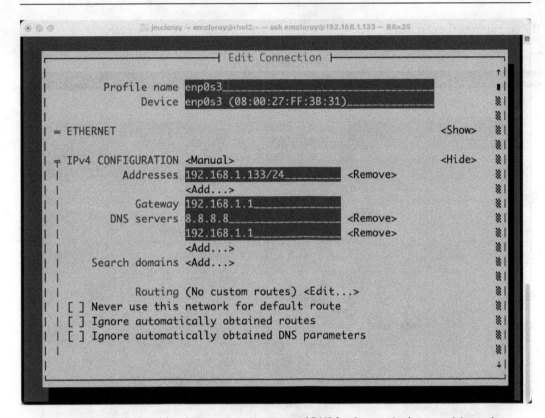

Figure 5.5 – IP address with subnet range, gateway, and DNS for the required connectivity options

From there, you would save this information and exit the nmtui GUI and then run the following command:

```
[emcleroy@rhel2 ~]$ sudo ifdown enp0s3 && sudo ifup enp0s3
```

The output from the command is shown in the following screenshot from the server:

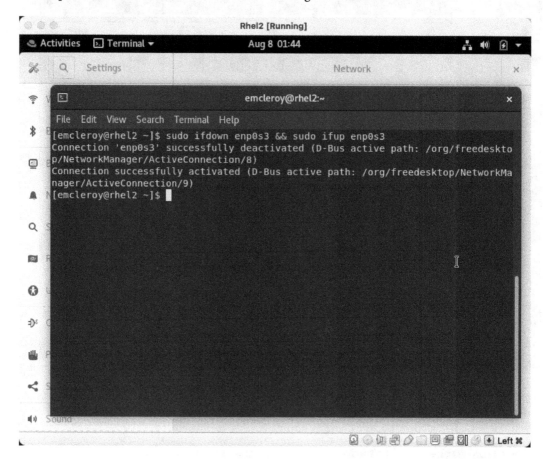

Figure 5.6 – Command to reload the interface to gain the new settings you specified

Alternatively, you can run the following command from SSH if you do not have direct access to the server:

```
[emcleroy@rhel2 ~]$ sudo systemctl restart NetworkManager.
service
```

In the following screenshot, you can see the output of the command:

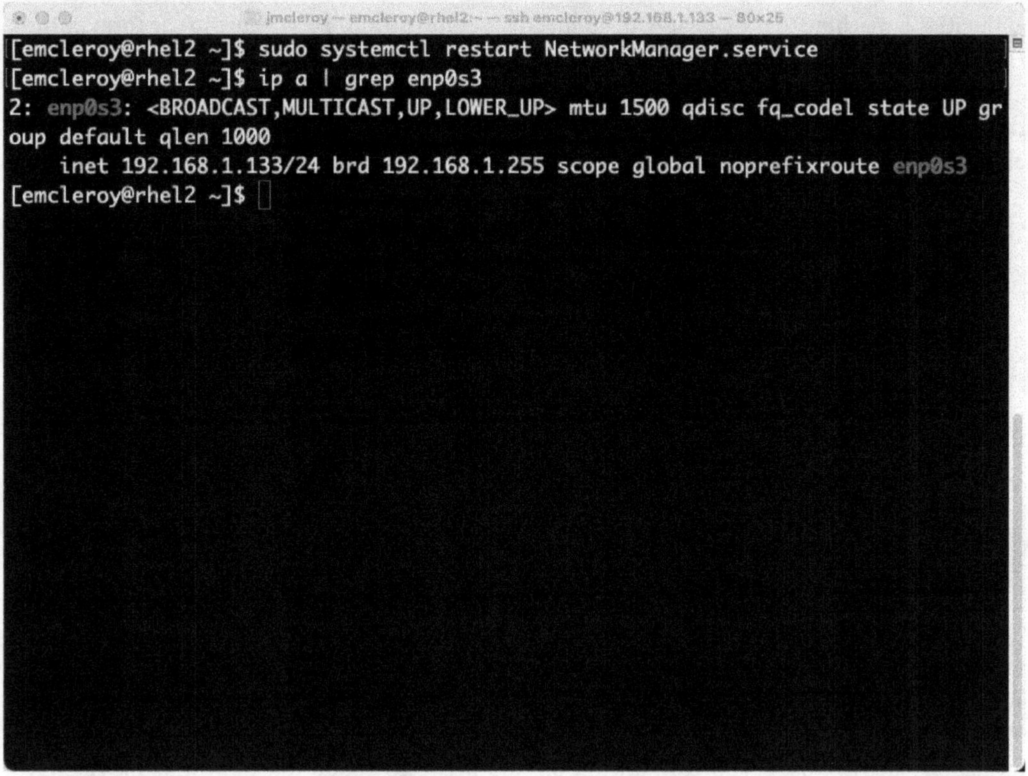
jmcleroy — emcleroy@rhel2:~ — ssh emcleroy@192.168.1.133 — 80x25
[emcleroy@rhel2 ~]$ sudo systemctl restart NetworkManager.service
[emcleroy@rhel2 ~]$ ip a | grep enp0s3
2: enp0s3: <BROADCAST,MULTICAST,UP,LOWER_UP> mtu 1500 qdisc fq_codel state UP group default qlen 1000
inet 192.168.1.133/24 brd 192.168.1.255 scope global noprefixroute enp0s3
[emcleroy@rhel2 ~]$

Figure 5.7 – Command to reload network manager in order to gain the new static settings

Using the basic out-of-the-box DHCP configuration to get online fast when available on your network

Out-of-the-box DHCP addressing will give you the IP addressing you need when physically connected to a network with a router and DHCP server running in order to obtain an IP automatically and connect to the network. This allows you to get online quickly without the need to know the connectivity settings such as subnet, gateway, or routing information.

Using DHCP at initial interface connectivity provided by an external source

When you first fire up your computer and plug it into the network, by default, most systems are configured to try to use DHCP. This handshake method allows for connectivity to the internet to occur faster and without direct intervention. For instance, in RHEL, when you fire up the interface when installing it, you will automatically have the configuration set to DHCP. This is due to most users having a serving DHCP server. This means that a handshake occurs between the DHCP server and the RHEL machine to provide it with connectivity information. This occurs in the form of the **DHCP Discover** message or the computer or client server asking for connectivity. The server replies with a **DHCP Offer** to provide the information. The client then responds with the **DHCP Request** for the information to connect. Finally, the DHCP server responds with a **DHCP Acknowledgement** that includes the information for connectivity. This can be seen in the following diagram:

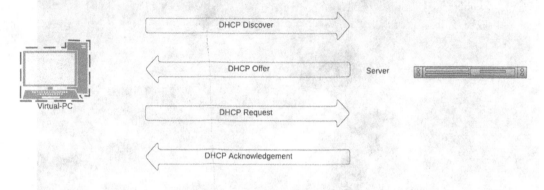

Figure 5.8 – DHCP handshake

Once you have the information from the DHCP server, your machine will be configured with an IP address, gateway, and DNS. This allows you to get online quicker and without the hassle of setting up the information yourself. In some cases, this is necessary because you neither have the required information (such as what IP addresses are free on your network) nor do you know what the gateway or DNS addresses are. By using DHCP, you are able to connect to the network reliably and without conflict and reach your local network.

In most enterprise environments, however, you will not use a DHCP server for application servers, but you will in most cases for providing connectivity to workers. Most application production servers are set through static or through automation using the next available application IP. This is why it is such a big deal to be able to set up a DHCP server as you will want the connectivity to be easy for your workers to be able to log in and get right to work. Knowing there is a need for a DHCP server, we will use it as a motivation to set up a DHCP server that will provide connectivity to internal workers at your company or your family at home.

With a DHCP server running and having connectivity, either through physical Ethernet or Wi-Fi, here is how you can set up DHCP IP addressing for your device. For this, you will need direct access to the server via the console or through the VM manager. Using the following command, you can check what your IP address is:

```
[emcleroy@rhel2 ~]$ ip a | grep enp0s3
```

As the following screenshot demonstrates, your device currently has no IP address:

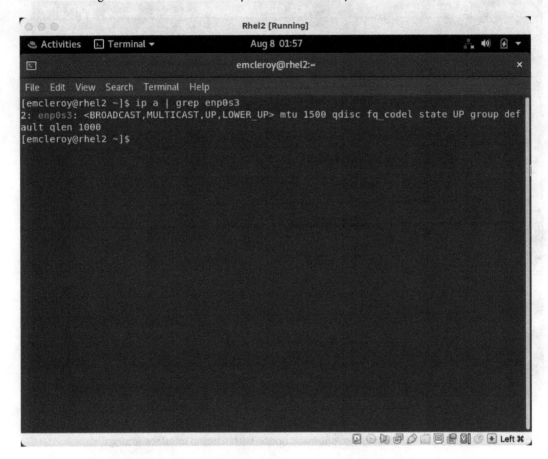

Figure 5.9 – There is currently no IP address associated with the connected enp0s3 interface

We are going to use `nmtui` again as we did when working with static addresses, but this time, we will choose **Automatic**, which utilizes DHCP, as shown in the following screenshot:

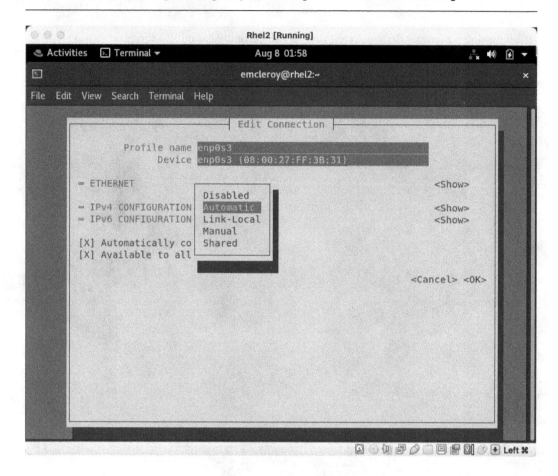

Figure 5.10 – Choosing the Automatic option within nmtui interface settings

After you choose the **Automatic** option and allow it to automatically configure your IP through DHCP, you will need to activate the interface so that it enables the connection. This is shown in the following screenshot from within nmtui:

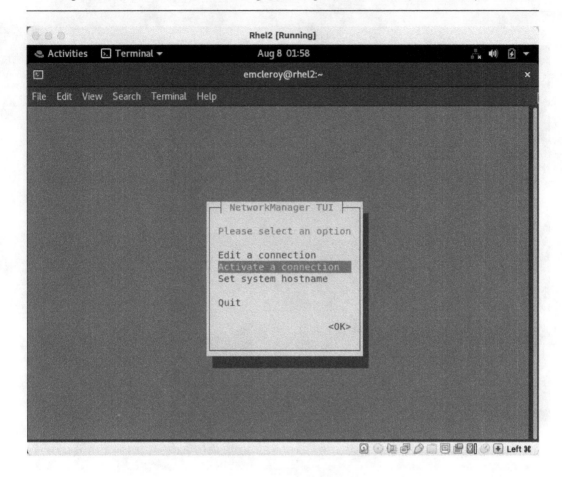

Figure 5.11 – Activating a connection within nmtui

After you choose **Activate a connection**, you then need to choose the interface that is physically connected, as shown in the following screenshot:

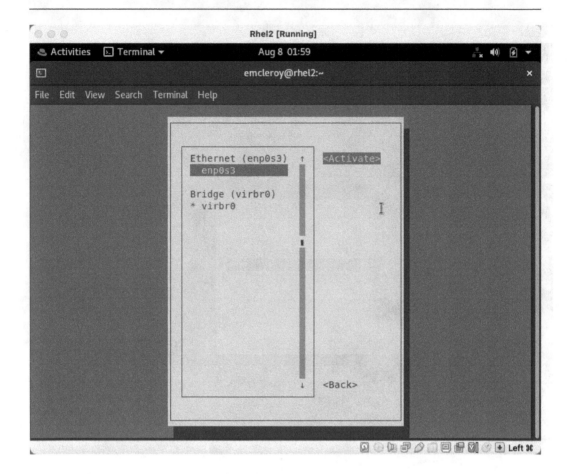

Figure 5.12 – Choosing the correct interface you previously set up to enable connectivity

After that is activated, the interface should come up and complete the DHCP handshake that was outlined previously to gain the information needed to make a network connection. Using the following command, you can see what IP address was obtained automatically:

```
[emcleroy@rhel2 ~]$ ip a | grep enp0s3
```

The connectivity information that was automatically obtained is shown in the following screenshot:

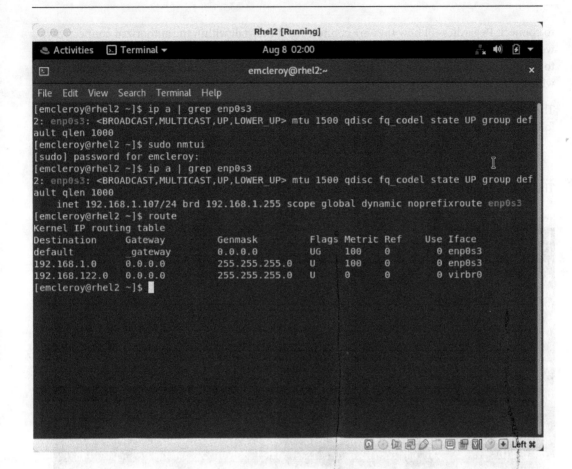

Figure 5.13 – The IP address and routing of the gateway

We have just set up your server or client machine to acquire network connectivity using DHCP. Through the automated system, you will be able to get online fast without having to worry about manually setting anything up. This is a big advantage for things that do not always need to have the same IP but only need network connectivity. Next, we will talk about creating a DHCP server to provide this type of functionality.

Setting up DHCP server configuration manually to provide DHCP services

In some cases, there will not be a DHCP server already set up on the network. As a system administrator, it may fall upon you to set one up. This is due to the need for a DHCP server for desktop support in many cases, and a necessity to get workers online fast and with minimal effort. This could mean that they are physically plugged into the network using Wi-Fi or a **Virtual Private Network (VPN)** to connect. Regardless of how they are connecting, if they fall within the authorized connectivity, we need to provide them with the correct addressing information. This allows them access not only to the internet but also to their systems within the intranet of the internal company. Let's get started with setting up a DHCP server manually.

The first thing you are going to need to do is to install the dhcp-server package using the following command:

```
[emcleroy@rhel1 ~]$ sudo dnf install dhcp-server
```

This output can be seen in the commands shown in the following screenshot:

```
[emcleroy@rhel1 ~]$ sudo dnf install dhcp-server
[sudo] password for emcleroy:
Updating Subscription Management repositories.
Red Hat Enterprise Linux 8 for x86_64 - AppStre 9.9 kB/s | 4.5 kB     00:00
Red Hat Enterprise Linux 8 for x86_64 - BaseOS   10 kB/s | 4.1 kB     00:00
Dependencies resolved.
========================================================================================
 Package          Arch    Version           Repository                         Size
========================================================================================
Installing:
 dhcp-server      x86_64  12:4.3.6-47.el8   rhel-8-for-x86_64-baseos-rpms 530 k
Upgrading:
 dhcp-client      x86_64  12:4.3.6-47.el8   rhel-8-for-x86_64-baseos-rpms 318 k
 bind-export-libs x86_64  32:9.11.36-3.el8  rhel-8-for-x86_64-baseos-rpms 1.1 M
 dhcp-common      noarch  12:4.3.6-47.el8   rhel-8-for-x86_64-baseos-rpms 207 k
 dhcp-libs        x86_64  12:4.3.6-47.el8   rhel-8-for-x86_64-baseos-rpms 148 k

Transaction Summary
========================================================================================
Install  1 Package
Upgrade  4 Packages

Total download size: 2.3 M
Is this ok [y/N]: 
```

Figure 5.14 – Installing the required dhcp-server package

From there, we will need to set the parameters within the correct configuration file so that dhcp-server knows what information to provide to the client machines. Using the following command, we can open and edit the dhcpd.conf file:

```
[emcleroy@rhel1 ~]$ sudo vi /etc/dhcp/dhcpd.conf
```

The command is shown in the following screenshot:

Figure 5.15 – Editing the dhcpd.conf file to set the required network settings

In order to save us somewhat of a headache, there is a trick that can be utilized. The trick is to look at the example of a configuration file and copy and paste the options. Once you've done that, you would then need to edit the options to match what your network requires to have the server provide the correct information to the clients. Using the following command, we can view the examples of working DHCP servers:

```
[emcleroy@rhel1 ~]$ cat /usr/share/doc/dhcp-server/dhcpd.conf.
example
```

This example can be found in the following screenshot:

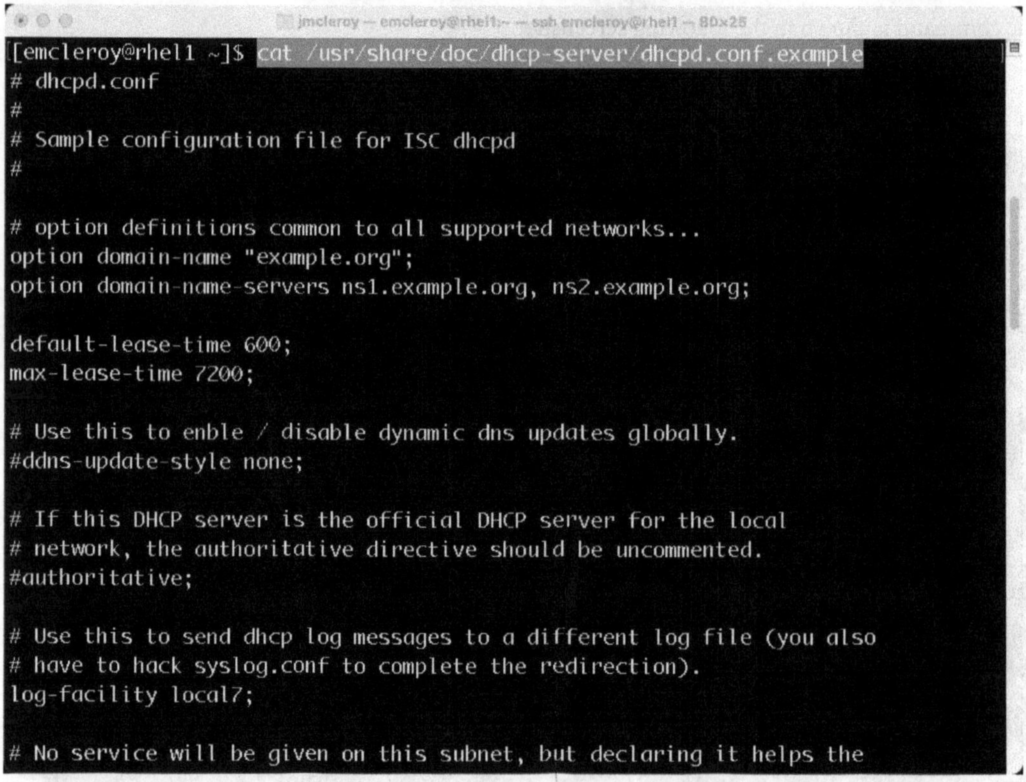

```
jmcleroy — emcleroy@rhel1:~ — ssh emcleroy@rhel1 — 80x25
[emcleroy@rhel1 ~]$ cat /usr/share/doc/dhcp-server/dhcpd.conf.example
# dhcpd.conf
#
# Sample configuration file for ISC dhcpd
#

# option definitions common to all supported networks...
option domain-name "example.org";
option domain-name-servers ns1.example.org, ns2.example.org;

default-lease-time 600;
max-lease-time 7200;

# Use this to enble / disable dynamic dns updates globally.
#ddns-update-style none;

# If this DHCP server is the official DHCP server for the local
# network, the authoritative directive should be uncommented.
#authoritative;

# Use this to send dhcp log messages to a different log file (you also
# have to hack syslog.conf to complete the redirection).
log-facility local7;

# No service will be given on this subnet, but declaring it helps the
```

Figure 5.16 – Example of a configured dhcpd.conf file

Further within the example, you will see the main portion of what is necessary to set up that internal DHCP server for your clients. This portion is shown in the following screenshot:

```
# A slightly different configuration for an internal subnet.
subnet 10.5.5.0 netmask 255.255.255.224 {
  range 10.5.5.26 10.5.5.30;
  option domain-name-servers ns1.internal.example.org;
  option domain-name "internal.example.org";
  option routers 10.5.5.1;
  option broadcast-address 10.5.5.31;
  default-lease-time 600;
  max-lease-time 7200;
}

# Hosts which require special configuration options can be listed in
# host statements.   If no address is specified, the address will be
# allocated dynamically (if possible), but the host-specific information
# will still come from the host declaration.

host passacaglia {
  hardware ethernet 0:0:c0:5d:bd:95;
  filename "vmunix.passacaglia";
  server-name "toccata.example.com";
}

# Fixed IP addresses can also be specified for hosts.   These addresses
# should not also be listed as being available for dynamic assignment.
# Hosts for which fixed IP addresses have been specified can boot using
```

Figure 5.17 – The necessary items fall under the "A slightly different configuration" section

By copying the subnet section and pasting it into the dhcpd.conf file, you can then edit it to the values you need easily without worrying about whether you missed any items. This configuration file is shown in the following screenshot:

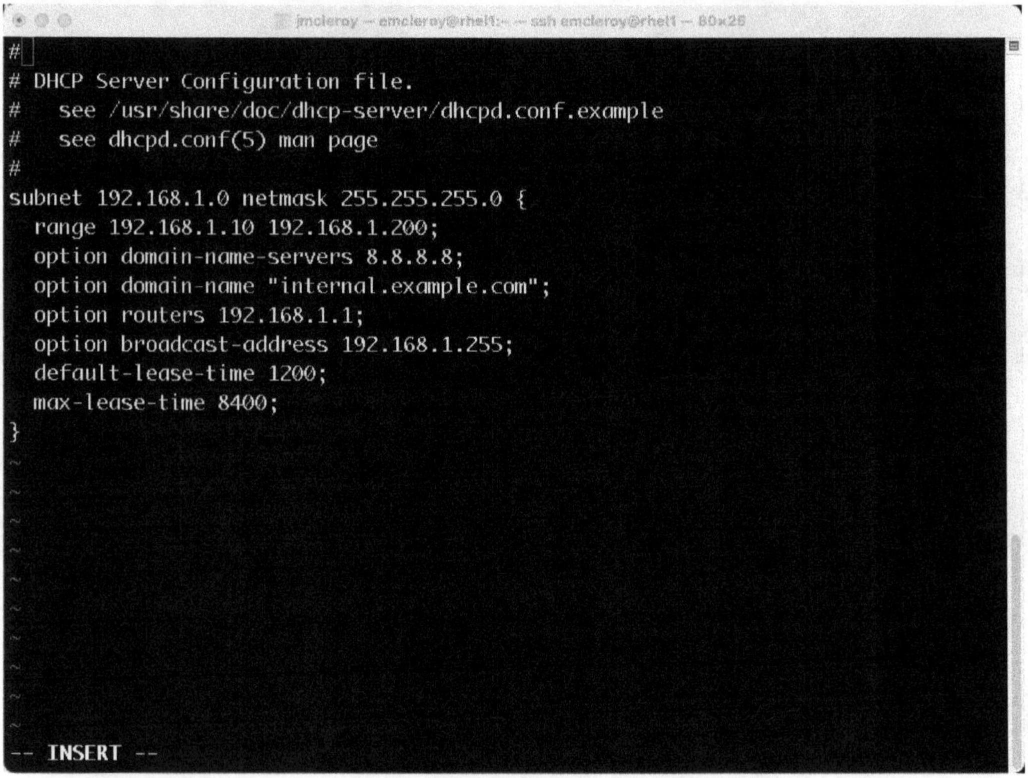

```
#
# DHCP Server Configuration file.
#   see /usr/share/doc/dhcp-server/dhcpd.conf.example
#   see dhcpd.conf(5) man page
#
subnet 192.168.1.0 netmask 255.255.255.0 {
  range 192.168.1.10 192.168.1.200;
  option domain-name-servers 8.8.8.8;
  option domain-name "internal.example.com";
  option routers 192.168.1.1;
  option broadcast-address 192.168.1.255;
  default-lease-time 1200;
  max-lease-time 8400;
}




-- INSERT --
```

Figure 5.18 – Configuration for dhcp-server that provides a 192.168.1.0/24 network

After you have all of the correct settings in your dhcpd.conf file, you can then start and enable the dhcp-server daemon in order to ensure that the server is running. We will use the following commands to enable and start the dhcpd service:

```
[emcleroy@rhel1 ~]$ sudo systemctl enable dhcpd
[emcleroy@rhel1 ~]$ sudo systemctl start dhcpd
```

The results of the commands are shown in the following screenshot:

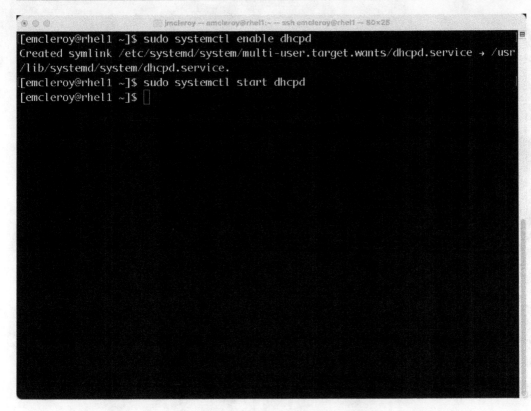

Figure 5.19 – Starting and enabling the service so that it starts at boot

After we have started and enabled the service, it is always a good idea to check and ensure that the system has successfully started. We can use the following command to view the status of the service:

```
[emcleroy@rhel1 ~]$ sudo systemctl status dhcpd
```

The status of running is shown in the following screenshot for dhcp-server:

```
[emcleroy@rhel1 ~]$ sudo systemctl status dhcpd
● dhcpd.service - DHCPv4 Server Daemon
   Loaded: loaded (/usr/lib/systemd/system/dhcpd.service; enabled; vendor prese>
   Active: active (running) since Mon 2022-08-08 01:55:07 EDT; 5min ago
     Docs: man:dhcpd(8)
           man:dhcpd.conf(5)
 Main PID: 3304 (dhcpd)
   Status: "Dispatching packets..."
    Tasks: 1 (limit: 11488)
   Memory: 5.2M
   CGroup: /system.slice/dhcpd.service
           └─3304 /usr/sbin/dhcpd -f -cf /etc/dhcp/dhcpd.conf -user dhcpd -grou>

Aug 08 01:57:24 rhel1.example.com dhcpd[3304]: reuse_lease: lease age 0 (secs) >
Aug 08 01:57:24 rhel1.example.com dhcpd[3304]: DHCPREQUEST for 192.168.1.106 (1>
Aug 08 01:57:24 rhel1.example.com dhcpd[3304]: DHCPACK on 192.168.1.106 to 74:a>
Aug 08 01:57:25 rhel1.example.com dhcpd[3304]: DHCPOFFER on 192.168.1.12 to 74:>
Aug 08 01:59:45 rhel1.example.com dhcpd[3304]: reuse_lease: lease age 265 (secs>
Aug 08 01:59:45 rhel1.example.com dhcpd[3304]: DHCPREQUEST for 192.168.1.73 fro>
Aug 08 01:59:45 rhel1.example.com dhcpd[3304]: DHCPACK on 192.168.1.73 to ee:f3>
Aug 08 01:59:45 rhel1.example.com dhcpd[3304]: reuse_lease: lease age 265 (secs>
Aug 08 01:59:45 rhel1.example.com dhcpd[3304]: DHCPREQUEST for 192.168.1.73 fro>
Aug 08 01:59:45 rhel1.example.com dhcpd[3304]: DHCPACK on 192.168.1.73 to ee:f3>
lines 1-22/22 (END)
```

Figure 5.20 – The status of running is shown for dhcp-server thus confirming it has been set up correctly

As with all other services, we want to reach the outside world, for which we need to open a firewall to allow incoming broadcast requests for a DHCP lease. We will use the following commands to open and reload the firewall:

```
[emcleroy@rhel1 ~]$ sudo firewall-cmd --permanent
--add-service-dhcp
[emcleroy@rhel1 ~]$ sudo firewall-cmd --reload
```

Opening the firewall command output is shown in the following screenshot:

```
[emcleroy@rhel1 ~]$ sudo firewall-cmd --permanent --add-service=dhcp
success
[emcleroy@rhel1 ~]$ sudo firewall-cmd --reload
success
[emcleroy@rhel1 ~]$
```

Figure 5.21 – Opening firewall ports for DHCP and reloading the firewall to allow connectivity

Previously, we skipped double-checking to ensure that the firewall rule was actually in place. This time, I wanted to showcase how you can check and see whether a service is allowed through your firewall. We will use the following command to view the services currently allowed by the firewall:

```
[emcleroy@rhel1 ~]$ sudo firewall-cmd --list-services
```

In the following screenshot, the command shows that the DHCP service is listed:

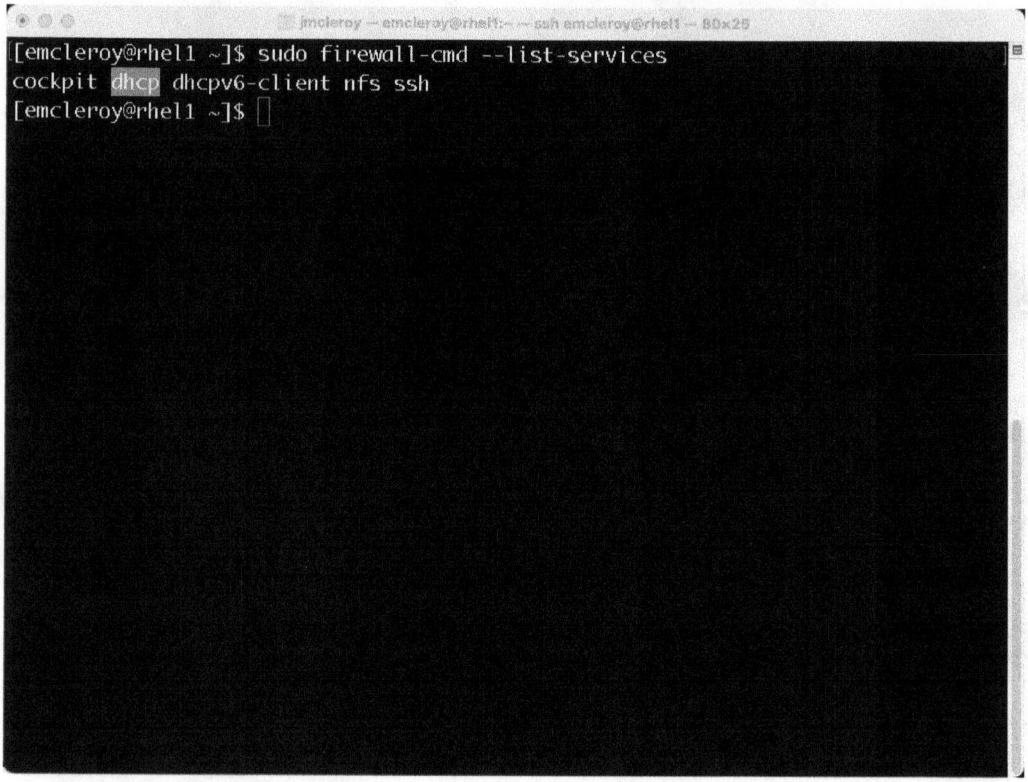

Figure 5.22 – Using the firewall-cmd --list-services command; you can see it includes DHCP

We have set up a DHCP server to provide connectivity to client devices connected to the subnet. This allows you to control what IP addresses devices receive from static DHCP addressing to next-in-line IP address pools. We briefly showcased options such as `default-lease-time`, which is the time the machine will keep the IP address before making a new request, for example. Also, with static MAC address mapping, it allows you to know ahead of time what IP address your machine will receive from a DHCP server. This is helpful when setting up Ansible playbooks to run against a specific IP address, for example. Next, we are going to talk about how to set this service up through Ansible Automation to ensure that it is easily repeatable and to lower the possibility of human error.

Automating DHCP server configuration to provide DHCP services

We are going to start by setting up Ansible's inventory of systems, which, in our case, is `rhel1. example.com` for our target device for the DHCP server. This will allow us to target only the device we are looking to run against; also, we will call out the server in our playbook to ensure that only `rhel1` is configured. Make a directory for your playbook, or you can have one directory with all of your playbooks, but it will use the same inventory so keep that in mind. In our case, we are going to make a directory in our `Documents` folder for the playbook and inventory to reside in, named `dhcp_server_playbook`:

```
[emcleroy@rhel3 Documents]$ mkdir dhcp_server_playbook
[emcleroy@rhel3 Documents]$ cd dhcp_server_playbook/
[emcleroy@rhel3 dhcp_server_playbook]$ ls
[emcleroy@rhel3 dhcp_server_playbook]$ vi inventory
```

The inventory file should look something like the following screenshot:

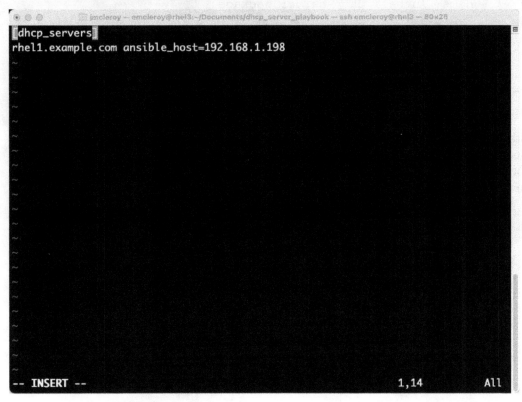

Figure 5.23 – DHCP server inventory file for the playbook to create the DHCP server

We will then move on to the playbook itself, where we will create tasks to install the DHCP server. First, we are going to want to point toward the inventory group we created so that no other hosts are updated. We will also set the `become` state to `true` in order to escalate our privileges, and `become_method` to `sudo` so that Ansible knows how to escalate privileges:

```
---
- name: Install and Configure DHCP Server
```

```
hosts: dhcp_servers
become: true
become_method: sudo
```

We then need to add some tasks to start building the DHCP server itself. We will start by installing the `dhcp-server` package using the `package` module. We will set the task to install the latest version of the package as well:

```
tasks:
  - name: Install dhcp-server package
    package:
      name: dhcp-server
      state: latest
```

After we have installed the server, we will then set the configuration files using the templating module through the use of a Jinja template.

First, let's look at the playbook layout, and then we will address the templates. We will use the `template` module to copy a template we have built to the system in order to configure `dhcp-server`. You will notice we are using a predefined variable of `playbook_dir` in order to point to the right location for the file. This is built into Ansible and is not something you need to configure yourself:

```
  - name: Set subnet configuration of DHCP-Server
    template:
      src: "{{ playbook_dir }}/templates/dhcpd.conf.j2"
      dest: /etc/dhcp/dhcpd.conf
```

Now, let's take a look at the `jinja2` template, which shows a normal `dhcpd.conf` layout for the subnet we want to create, as shown in the following screenshot:

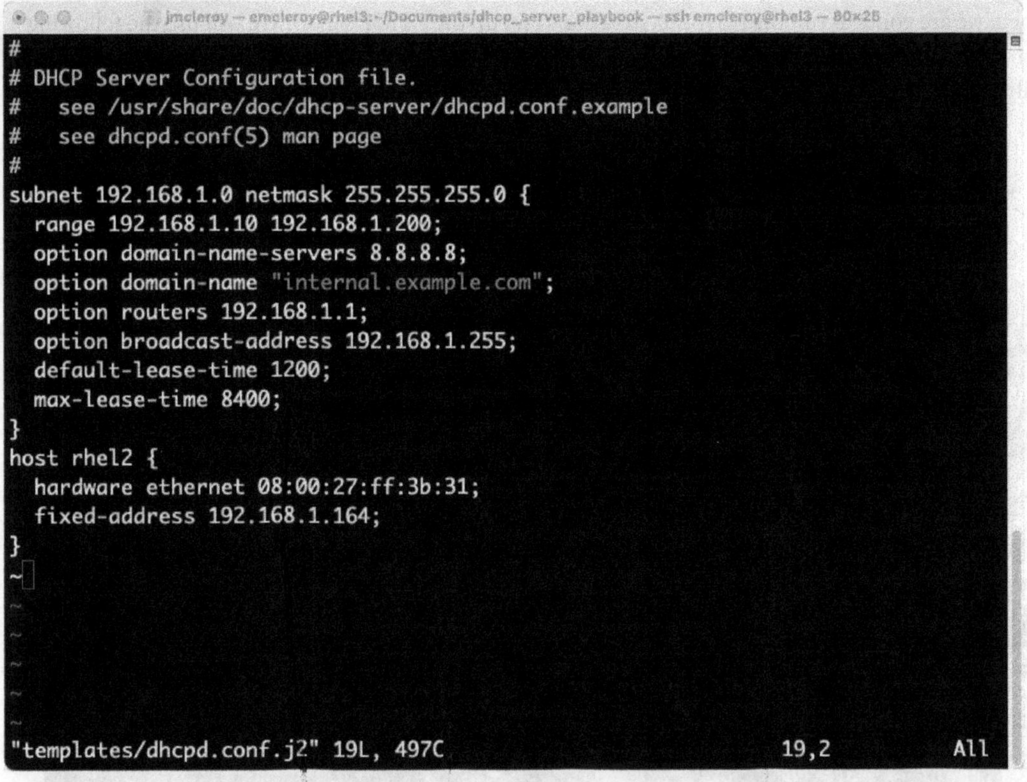

```
#
# DHCP Server Configuration file.
#   see /usr/share/doc/dhcp-server/dhcpd.conf.example
#   see dhcpd.conf(5) man page
#
subnet 192.168.1.0 netmask 255.255.255.0 {
  range 192.168.1.10 192.168.1.200;
  option domain-name-servers 8.8.8.8;
  option domain-name "internal.example.com";
  option routers 192.168.1.1;
  option broadcast-address 192.168.1.255;
  default-lease-time 1200;
  max-lease-time 8400;
}
host rhel2 {
  hardware ethernet 08:00:27:ff:3b:31;
  fixed-address 192.168.1.164;
}
~
~
~
~
~
"templates/dhcpd.conf.j2" 19L, 497C                          19,2          All
```

Figure 5.24 – dhcpd.conf.j2 file for use with the Ansible playbook

After this is completed, we will then start and enable the service. Open the firewall rules for dhcp and reload the firewall:

```
    - name: Enable and start dhcp-server service
      service:
        name: dhcpd
        enabled: true
        state: restarted

    - name: Open firewall rules
      firewalld:
        permanent: true
        immediate: true
        service: dhcp
        state: enabled
```

Then, we are going to run the playbook in order to set up the DHCP server. We will use the following command; if you have keys fully set up, this is not necessary, but I wanted to show you how to do it with passwords:

```
[emcleroy@rhel3 dhcp_server_playbook]$ ansible-playbook -i
inventory dhcp_server_playbook.yml -u emcleroy -k --ask-become
```

The output from running the playbook is shown in the following screenshot:

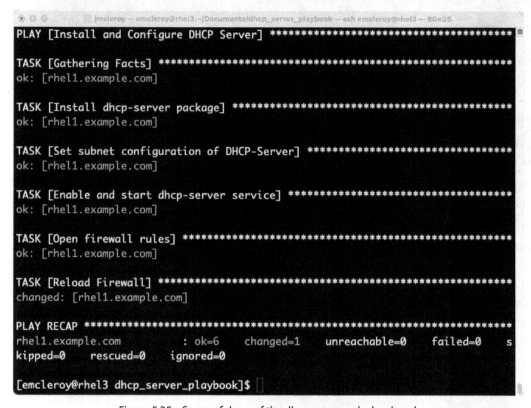

Figure 5.25 – Successful run of the dhcp_server_playbook.yml

With this running successfully, you can point your server toward the system to look for a DNS server. Since my system is currently on a DHCP server, it makes it harder to test as I am bridging the connections, but you can change the configuration to use a non-bridged connection such as an internal network in VirtualBox and you will be able to test the DHCP to see what it provides.

In this section, we learned how to automate the setup of a DHCP server. This allows us to provide DHCP to systems automatically. This allows the ease of use of connecting to a new network for client computers when running in an enterprise environment. Next, we will discuss the DNS and how that relates to network connectivity.

Learning about the DNS and why you need to know about it

DNS is important to learn about as it simplifies how we find addresses of networked devices on the world wide web. Without the DNS to allow FQDNs to translate into IP addresses, we would all have to search for websites via IP. This would become confusing as there are so many websites we use today that remembering all the IP addresses for them would be hectic and unmanageable. This would inevitably lead to a database that you would have to look at each time you wanted to go to a website. Thus, the DNS was born, and it provides that function allowing you to remember friendly names of websites in order to access them, such as `google.com`.

Setting up DNS server configuration manually to provide DNS services

The first thing we will need to do is install the `bind` package for the DNS to be able to run. This allows for the daemon needed to serve the DNS entries to the systems. We will start by using `sudo dnf install bind -y`, with the output shown in the following screenshot:

```
Updating Subscription Management repositories.
Last metadata expiration check: 0:00:10 ago on Mon 08 Aug 2022 03:07:03 AM EDT.
Dependencies resolved.
================================================================================
 Package          Arch    Version            Repository                    Size
================================================================================
Installing:
 bind             x86_64  32:9.11.36-3.el8   rhel-8-for-x86_64-appstream-rpms 2.1 M
Upgrading:
 bind-libs        x86_64  32:9.11.36-3.el8   rhel-8-for-x86_64-appstream-rpms 175 k
 python3-bind     noarch  32:9.11.36-3.el8   rhel-8-for-x86_64-appstream-rpms 150 k
 bind-libs-lite   x86_64  32:9.11.36-3.el8   rhel-8-for-x86_64-appstream-rpms 1.2 M
 bind-utils       x86_64  32:9.11.36-3.el8   rhel-8-for-x86_64-appstream-rpms 451 k
 bind-license     noarch  32:9.11.36-3.el8   rhel-8-for-x86_64-appstream-rpms 103 k
 json-c           x86_64  0.13.1-3.el8       rhel-8-for-x86_64-baseos-rpms    41 k
Installing dependencies:
 fstrm            x86_64  0.6.1-2.el8        rhel-8-for-x86_64-appstream-rpms  29 k

Transaction Summary
================================================================================
Install  2 Packages
Upgrade  6 Packages

Total download size: 4.2 M
Is this ok [y/N]:
```

Figure 5.26 – Bind being installed through the package manager

After `bind` is installed, we are going to go to the documentation to get a clear look at the items that need to be set. We can find the `named.conf` file example shown at the following path in the screenshot:

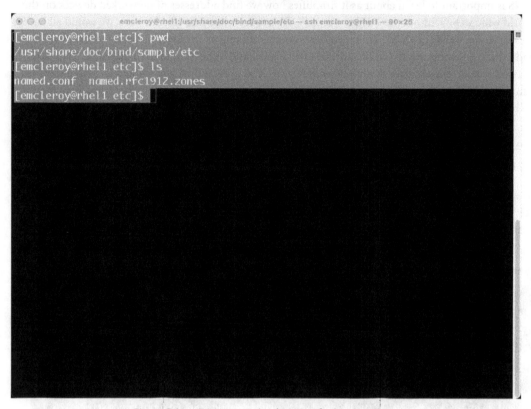

Figure 5.27 – Documentation location for bind settings

Once we have installed `bind` for the DNS, we are going to configure the `/etc/named.conf` file in order to set up the needed parameters. This file's location is shown in the following screenshot and edited with the `sudo vi /etc/named.conf` command:

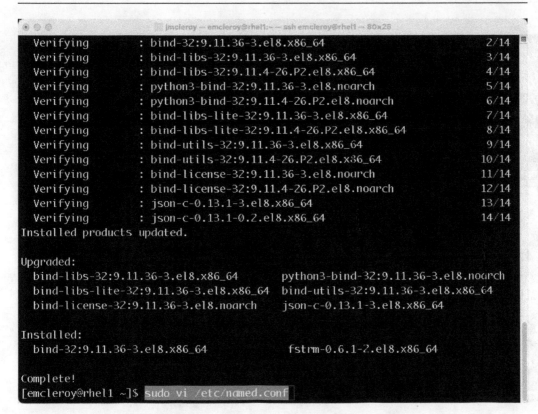

Figure 5.28 – Command to edit the /etc/named.conf file to configure the DNS

You will notice that the first things that need to be set up are the listening ports and interfaces. The DNS listens on port 53 as that is the well-known (default) port for the DNS that is set up in the networking world. We are going to adjust the settings so that we listen for incoming requests on port 53 on any interface connected to the network. This configuration is shown in the following screenshot:

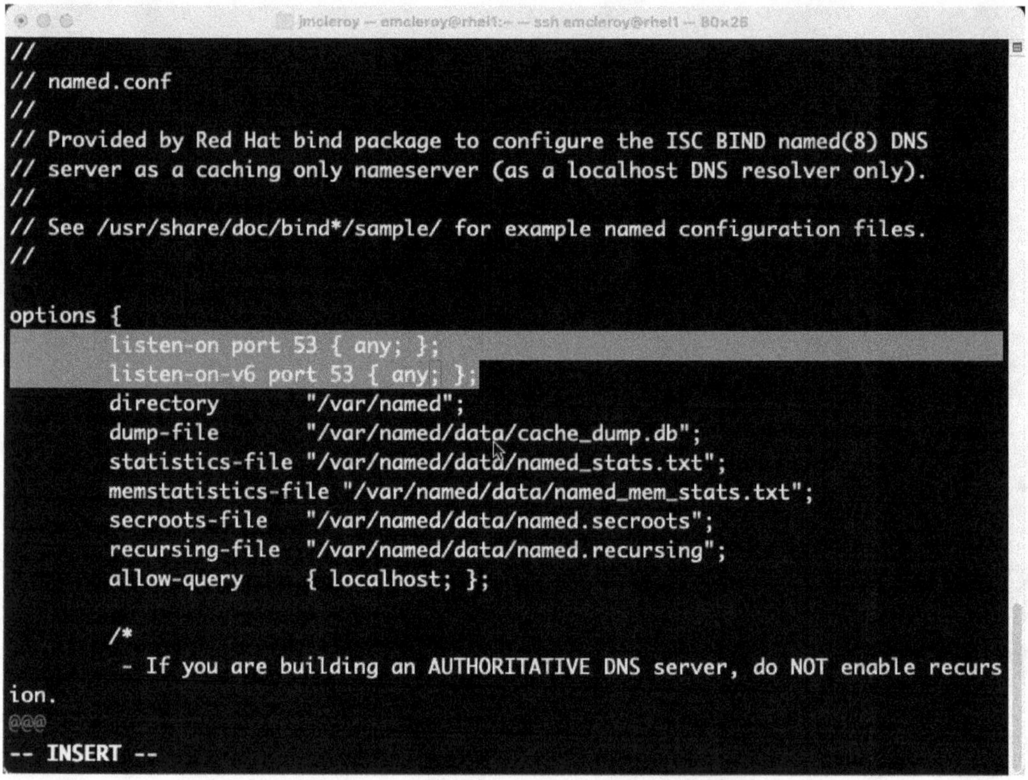

```
//
// named.conf
//
// Provided by Red Hat bind package to configure the ISC BIND named(8) DNS
// server as a caching only nameserver (as a localhost DNS resolver only).
//
// See /usr/share/doc/bind*/sample/ for example named configuration files.
//

options {
        listen-on port 53 { any; };
        listen-on-v6 port 53 { any; };
        directory       "/var/named";
        dump-file       "/var/named/data/cache_dump.db";
        statistics-file "/var/named/data/named_stats.txt";
        memstatistics-file "/var/named/data/named_mem_stats.txt";
        secroots-file   "/var/named/data/named.secroots";
        recursing-file  "/var/named/data/named.recursing";
        allow-query     { localhost; };

        /*
         - If you are building an AUTHORITATIVE DNS server, do NOT enable recurs
ion.
@@@
-- INSERT --
```

Figure 5.29 – Interface settings for DNS port 53

The next portion of the configuration file we are concerned with is the location of the zone files. A zone file is what tells the DNS what IP addresses to give to an FQDN (these files are what are looked up when you search for a website such as www.google.com), which then returns an IP address to your computer via the DNS so that it can reach the destination. In this case, as you can see in the following screenshot, the location of these files is the /var/named directory:

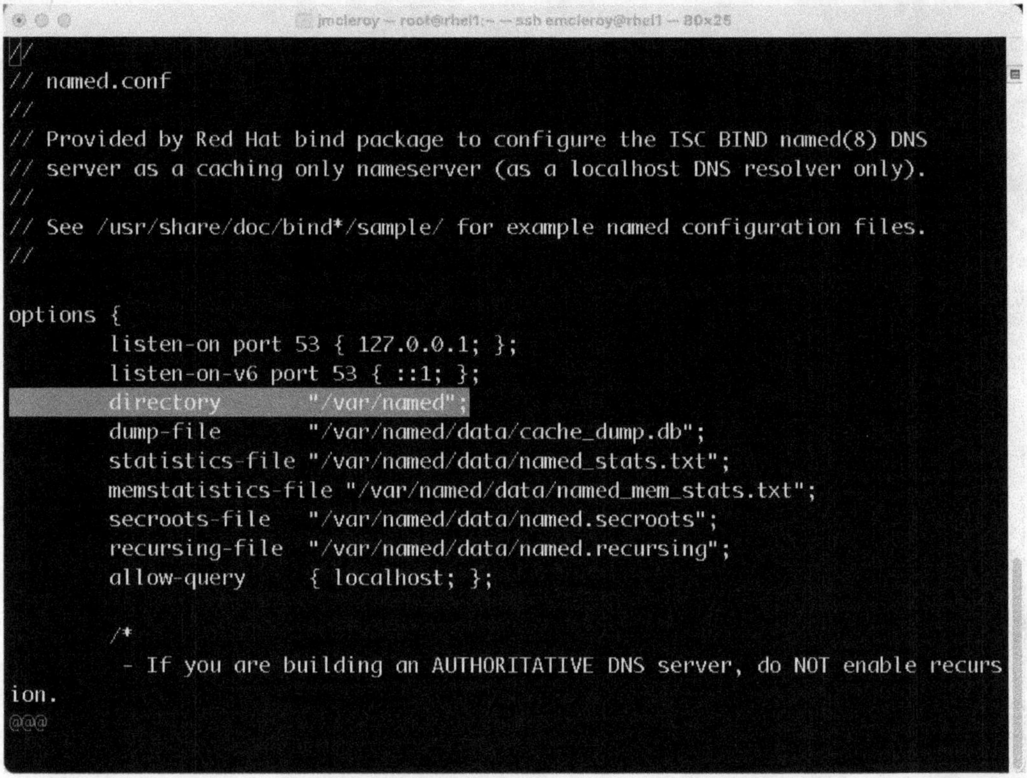
jmcleroy — root@rhel1:~ — ssh emcleroy@rhel1 — 80x25
//
// named.conf
//
// Provided by Red Hat bind package to configure the ISC BIND named(8) DNS
// server as a caching only nameserver (as a localhost DNS resolver only).
//
// See /usr/share/doc/bind*/sample/ for example named configuration files.
//
options {
listen-on port 53 { 127.0.0.1; };
listen-on-v6 port 53 { ::1; };
directory "/var/named";
dump-file "/var/named/data/cache_dump.db";
statistics-file "/var/named/data/named_stats.txt";
memstatistics-file "/var/named/data/named_mem_stats.txt";
secroots-file "/var/named/data/named.secroots";
recursing-file "/var/named/data/named.recursing";
allow-query { localhost; };
/*
- If you are building an AUTHORITATIVE DNS server, do NOT enable recurs
ion.
@@@

Figure 5.30 – Zone file directory location

When setting up an authoritative DNS server, or what is known as a top-level DNS server, you will want to set `recursion` to `no` for something like a lab environment or even an exam where you would like DNS to learn from other DNS servers' upstream information. Then, to save that information, such as a DNS entry, you will want `recursion` set to `yes`. In our case, for this example, we are setting it to `no` as we want to be top level, as shown in the following screenshot:

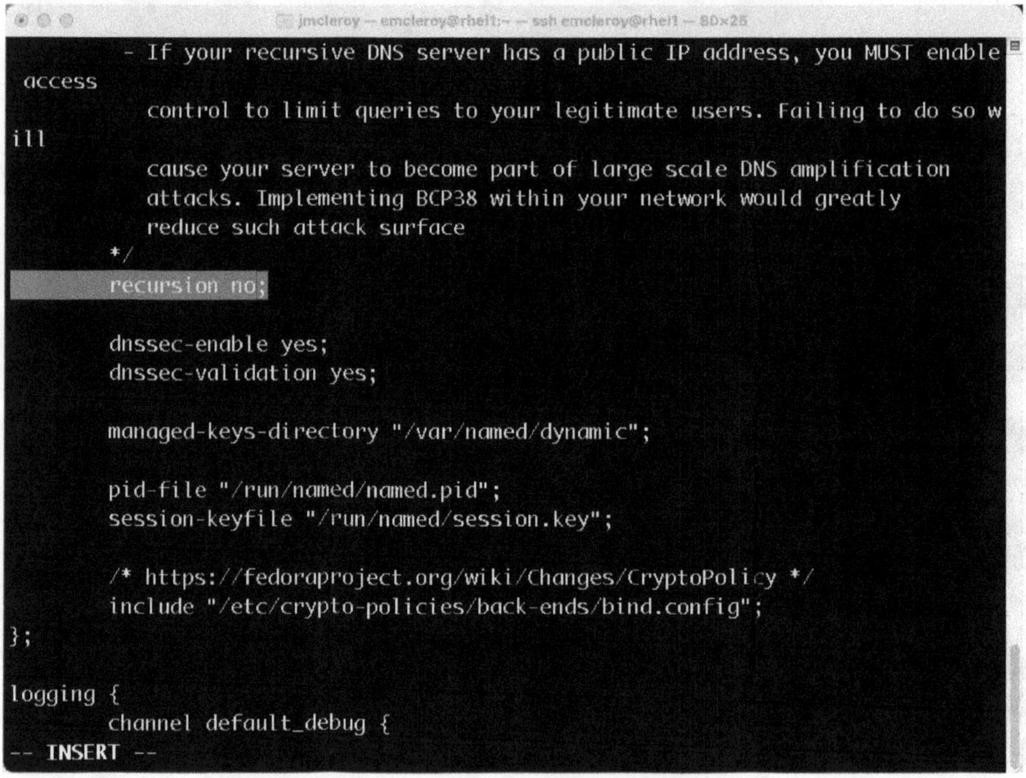

Figure 5.31 – Recursion is set to no for authoritative DNS servers

The next thing we are going to set up is a zone file location for a master forward lookup zone. A forward lookup zone is used to look up the IP address of the DNS FQDN when provided. This allows for your computer or server to be able to find the item that it is attempting to reach over the internet or intranet. The example.com zone is shown in the following screenshot:

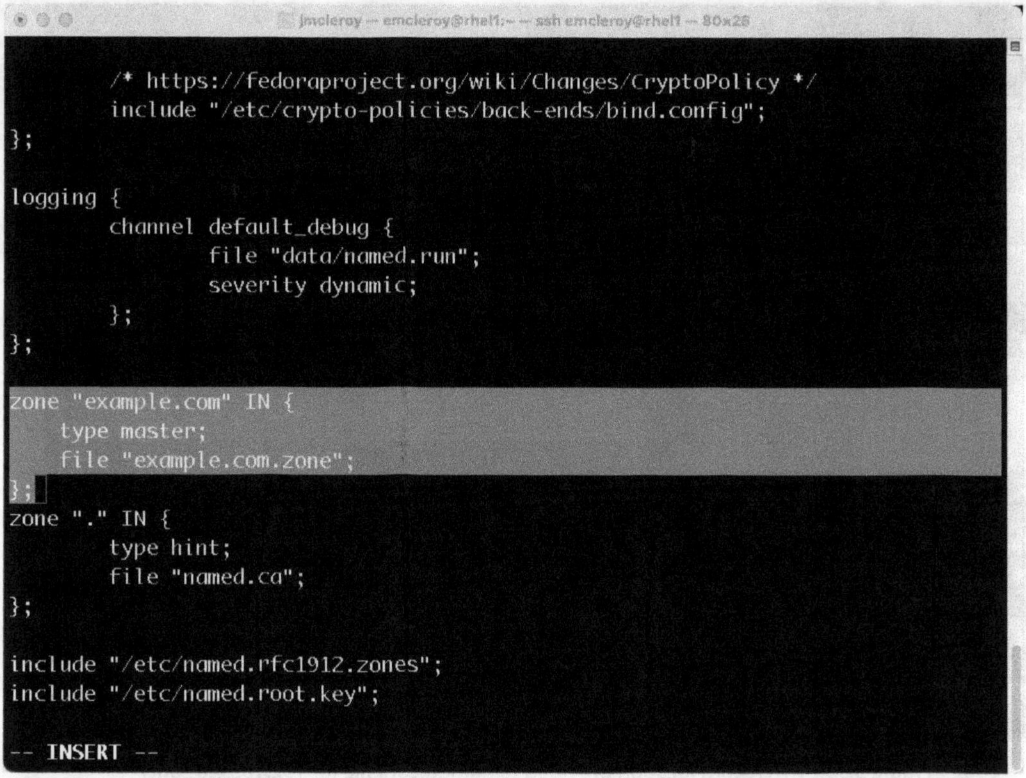
jmcleroy — emcleroy@rhel1:~ — ssh emcleroy@rhel1 — 80x28
/* https://fedoraproject.org/wiki/Changes/CryptoPolicy */
include "/etc/crypto-policies/back-ends/bind.config";
};
logging {
channel default_debug {
file "data/named.run";
severity dynamic;
};
};
zone "example.com" IN {
type master;
file "example.com.zone";
};
zone "." IN {
type hint;
file "named.ca";
};
include "/etc/named.rfc1912.zones";
include "/etc/named.root.key";
-- INSERT --

Figure 5.32 – The example.com zone file location provided to the named.conf file

Next, we are going to set up a reverse zone lookup file location within the `named.conf` file. This is so that if you were to look up the IP address in your system, it would provide the FQDN for that IP address. This **Pointer Record** (**PTR**) allows for these types of lookups to occur and is necessary for things such as SSL certificates. In the following screenshot, we show you the setting up of the `192.168.1.0` PTR record, which, in our case, will point to `example.com`:

```
logging {
        channel default_debug {
                file "data/named.run";
                severity dynamic;
        };
};

zone "example.com" IN {
        type master;
        file "example.com.zone";
};

zone "1.168.192.in-addr.arpa" IN {
        type master;
        file "192.168.1.zone";
};

zone "." IN {
        type hint;
        file "named.ca";
};

include "/etc/named.rfc1912.zones";
-- INSERT --
```

Figure 5.33 – PTR record location provided to named.conf file for 192.168.1.0 zone

Now that we have added the needed zone file locations to named.conf, we need to create the zone files for DNS to be able to look up information related to the DNS records. We can use the localhost example provided in /var/named/named.localhost to showcase the needed items for a zone file. This zone file example is shown in the following screenshot:

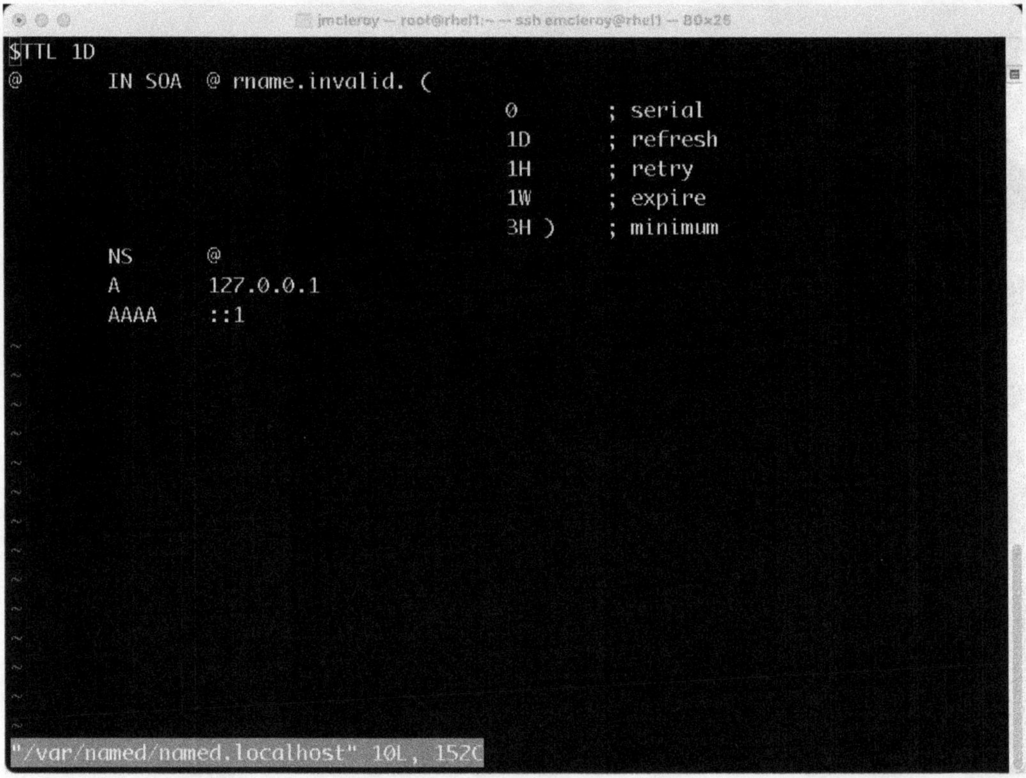

```
$TTL 1D
@       IN SOA  @ rname.invalid. (
                                    0           ; serial
                                    1D          ; refresh
                                    1H          ; retry
                                    1W          ; expire
                                    3H )        ; minimum
        NS      @
        A       127.0.0.1
        AAAA    ::1
~
~
~
~
~
~
~
~
~
~
"/var/named/named.localhost" 10L, 152C
```

Figure 5.34 – Example zone file for DNS records

As shown in the previous screenshot, there are some different items we need to go into detail about. First would be the **Time to Live** (**TTL**), which is how long a record will cache on a system before it needs to be refreshed. The next item is the **Start of Authority** (**SOA**), which dictates the primary authoritative server for the DNS records to be served from. After that, you will then add the email address (replacing the @ with a .) of the administrator of the DNS server. The next important items listed in this file are the records themselves. These include A records (which are IPV4 DNS records), AAAA records (which are IPV6 records), and NS records (which are name server records and determine what the DNS servers that are utilized are set at and are needed for all zone files). The serial number is important as it helps to identify whether a DNS record that is retrieved by the host machine is up to date or not. The end . (dot) is used on the named items when they are FQDNs that can be reached. In the following screenshot, you will see how we set up a simple zone file for example.com:

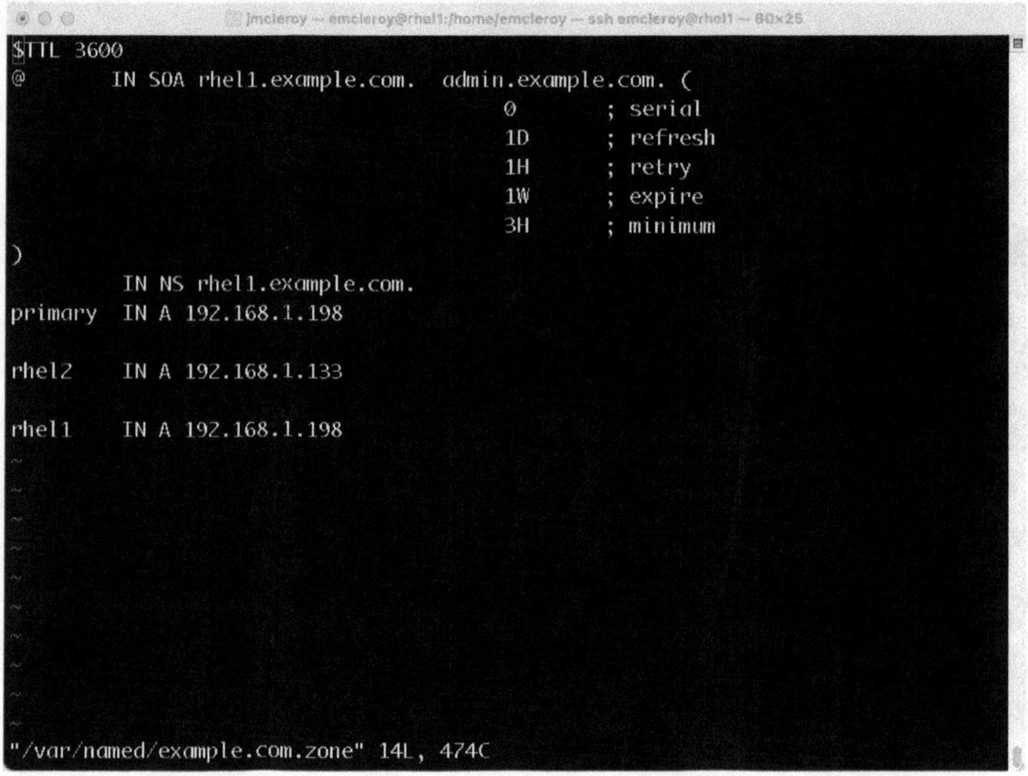

```
$TTL 3600
@        IN SOA rhel1.example.com.  admin.example.com. (
                                    0          ; serial
                                    1D         ; refresh
                                    1H         ; retry
                                    1W         ; expire
                                    3H         ; minimum
)
         IN NS rhel1.example.com.
primary  IN A 192.168.1.198

rhel2    IN A 192.168.1.133

rhel1    IN A 192.168.1.198
~
~
~
~
~
~
~
~
~
~
~
"/var/named/example.com.zone" 14L, 474C
```

Figure 5.35 – Zone file configured for example.com with A records and NS records set

Following that zone setup, we will then set up the PTR zone for reverse DNS lookup. This is accomplished in the same format but with the reverse DNS layout of the IP address in reverse of the record, which allows servers to be discovered using the IP address instead of the name. This record type and setup are shown in the following screenshot:

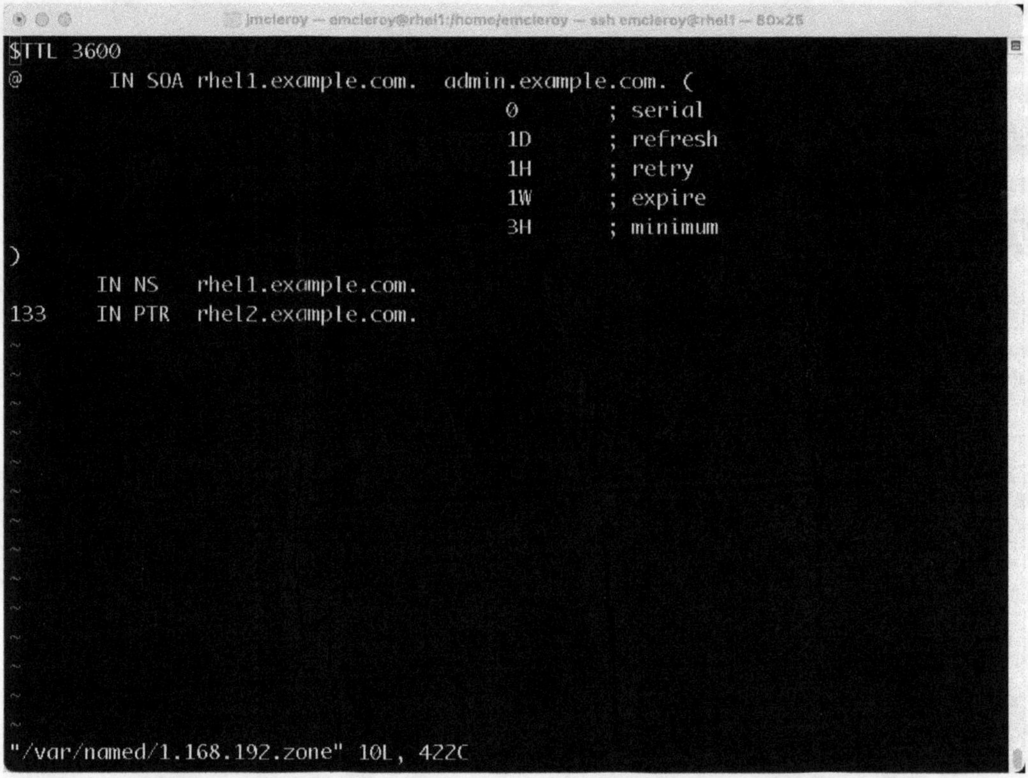

```
$TTL 3600
@       IN SOA rhel1.example.com.   admin.example.com. (
                                    0           ; serial
                                    1D          ; refresh
                                    1H          ; retry
                                    1W          ; expire
                                    3H          ; minimum
)
        IN NS    rhel1.example.com.
133     IN PTR   rhel2.example.com.
~
~
~
~
~
~
~
~
~
~
~
~
~
~
"/var/named/1.168.192.zone" 10L, 422C
```

Figure 5.36 – Zone file for PTR reverse DNS lookup of the 192.168.1.0 subnet for example.com

After our configuration files are set up and our zone files are populated, we will then want to enable and start the named service provided by the bind package to provide the DNS to systems pointed at it for DNS records. This is accomplished with the commands in the following screenshot:

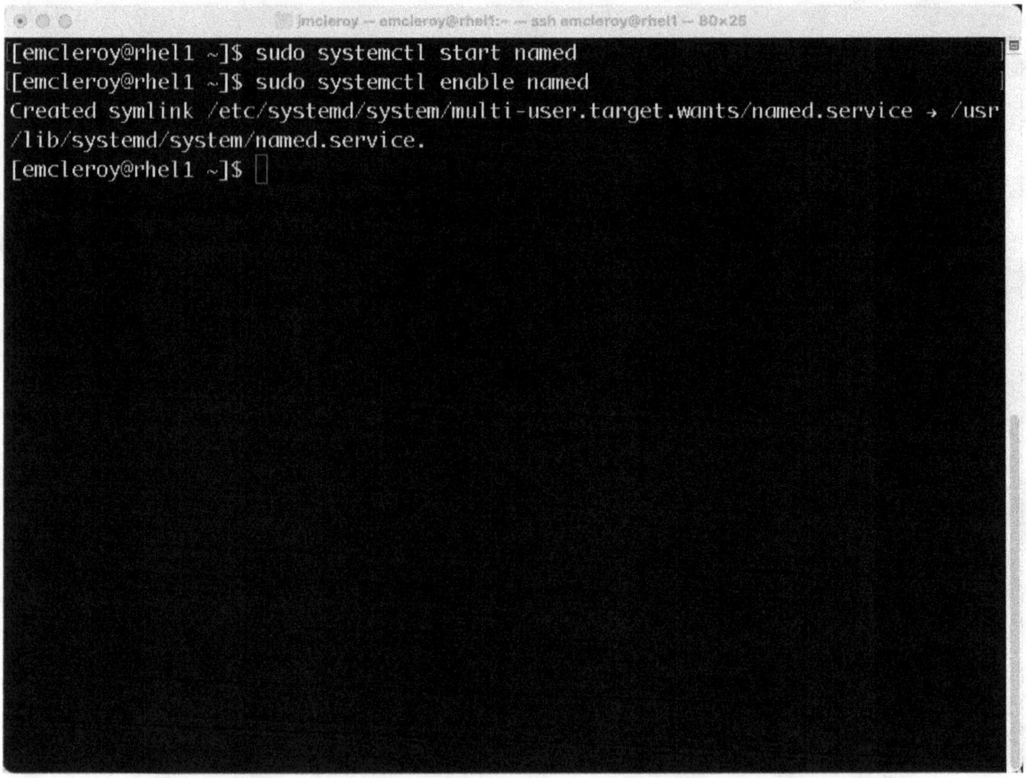

```
[emcleroy@rhel1 ~]$ sudo systemctl start named
[emcleroy@rhel1 ~]$ sudo systemctl enable named
Created symlink /etc/systemd/system/multi-user.target.wants/named.service → /usr
/lib/systemd/system/named.service.
[emcleroy@rhel1 ~]$
```

Figure 5.37 – Start and enable the named service to provide the DNS

To ensure that the DNS is up and running, we will use the following command:

```
[emcleroy@rhel1 ~]$ sudo systemctl status named
```

We can now view the output status of the named service, as shown in the following screenshot, which shows it is actively running without any failures:

Figure 5.38 – Status of the named service for the DNS

In order to ensure that the DNS service can be reached by outside systems, we need to allow it within the firewall. We will use the following commands to start, enable, and apply firewall rules:

```
[emcleroy@rhel1 ~]$ sudo systemctl start named
[emcleroy@rhel1 ~]$ sudo systemctl enable named
[emcleroy@rhel1 ~]$ sudo firewall-cmd --permanent
--add-service-dns
[emcleroy@rhel1 ~]$ sudo firewall-cmd --reload
```

The firewalld setup, start and enable output, and commands are shown in the following screenshot:

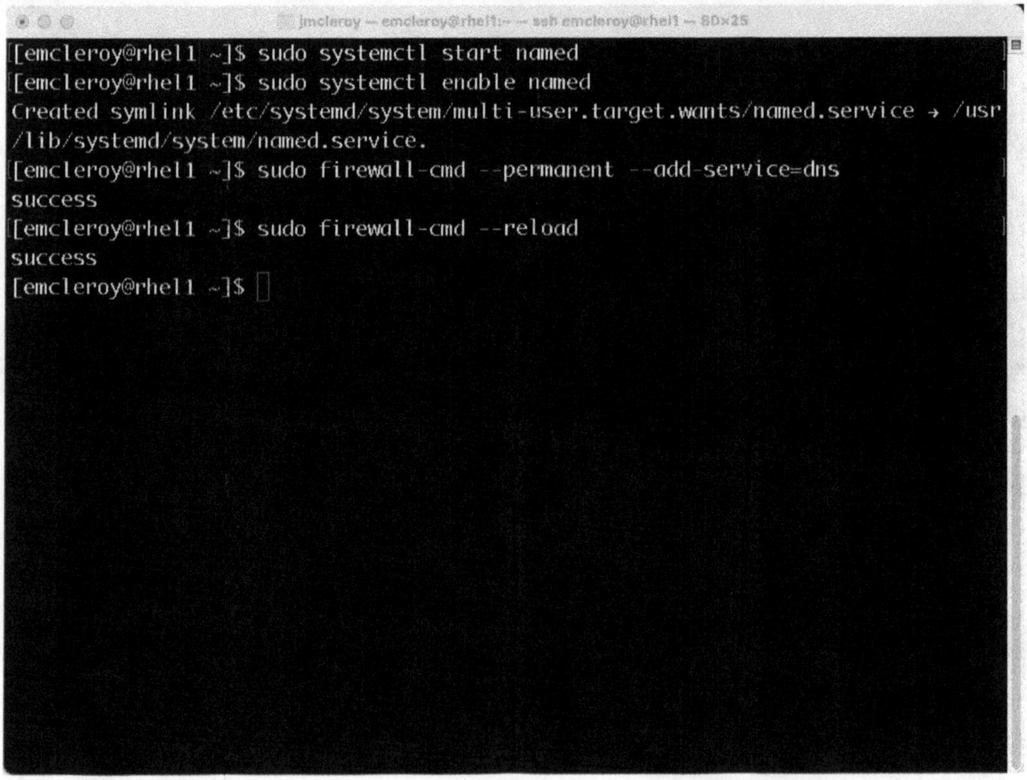

```
[emcleroy@rhel1 ~]$ sudo systemctl start named
[emcleroy@rhel1 ~]$ sudo systemctl enable named
Created symlink /etc/systemd/system/multi-user.target.wants/named.service → /usr
/lib/systemd/system/named.service.
[emcleroy@rhel1 ~]$ sudo firewall-cmd --permanent --add-service=dns
success
[emcleroy@rhel1 ~]$ sudo firewall-cmd --reload
success
[emcleroy@rhel1 ~]$
```

Figure 5.39 – Allowed DNS and then reloaded the firewall for connectivity

To ensure that the system is providing the DNS records that we set up in the zone files, we will do a dig lookup on the specific DNS server for a system we propagated in the zone file. Keep in mind that whenever you update a zone file, you need to reload named.service as well. We will use the dig command to do a DNS lookup:

```
[emcleroy@rhel1 ~]$ dig rhel2.example.com @192.168.1.198
```

The following screenshot shows the dig or lookup of a record from a DNS server from the DNS server we created the information for rhel2.example.com:

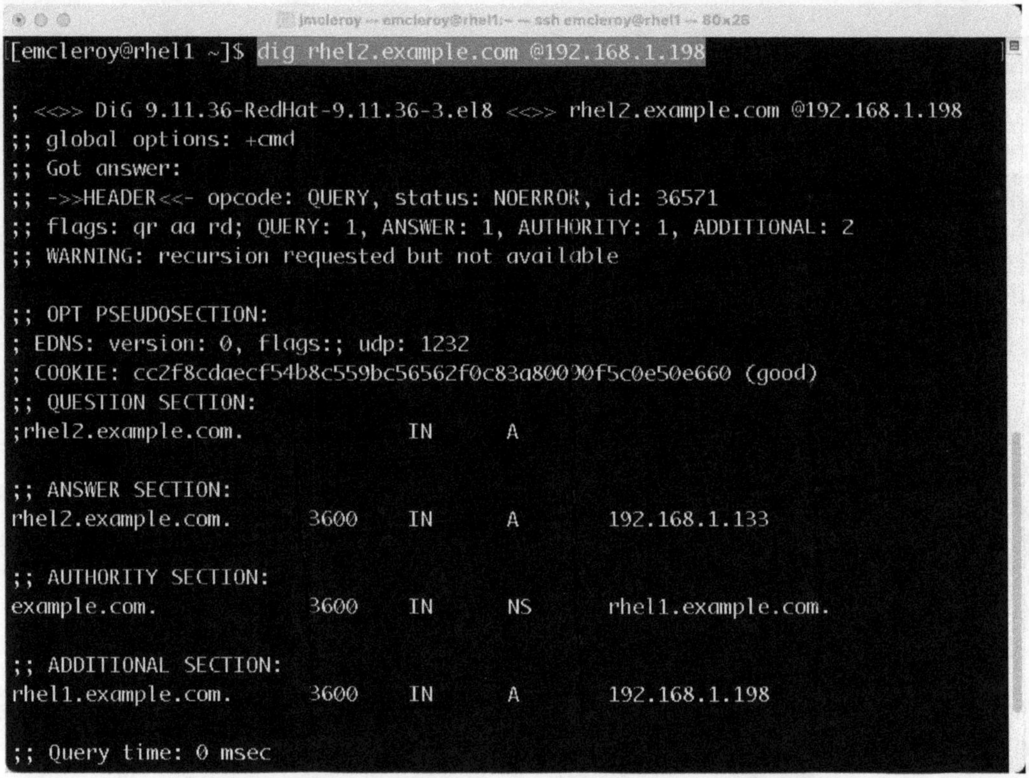

Figure 5.40 – DNS dig lookup of the rhel2.example.com server

In this section, we learned how to manually set up a DNS server. This is crucial in enhancing your understanding of how the DNS works and why it is used. We showed how forward and reverse zones are set up to allow for connectivity as well as installing the service needed. Next, we are going to delve into automating a DNS server creation using Ansible Automation.

Automating DNS server configuration to provide DNS services

Next up, we are going to learn how to set up a DNS server using Ansible Automation through Jinja templates as we did with our DHCP setup. These files might need to be altered for your setup, but you will come to understand the principles of how to complete the Ansible playbook.

We are going to start by creating the directory for the playbook using the following commands:

```
[emcleroy@rhel1 ~]$ ls
[emcleroy@rhel1 ~]$ cd Documents/
[emcleroy@rhel1 ~]$ mkdir dns_server
```

The output of these commands can be seen in the following screenshot:

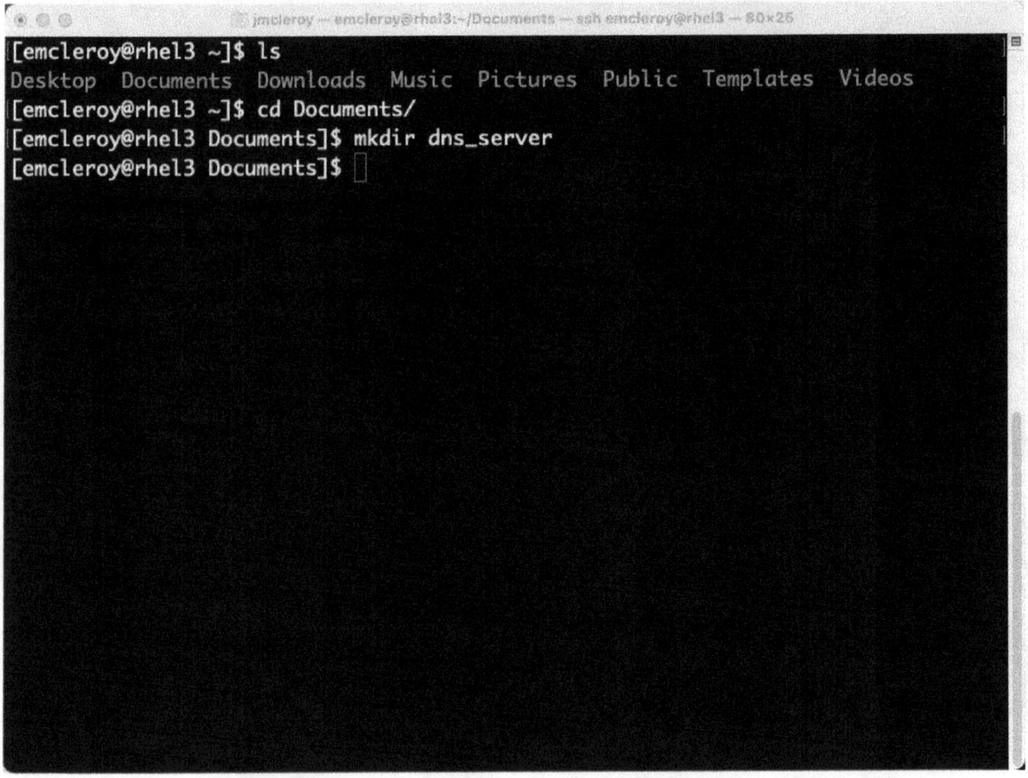
jmcleroy — emcleroy@rhel3:~/Documents — ssh emcleroy@rhel3 — 80×25
[emcleroy@rhel3 ~]$ ls
Desktop Documents Downloads Music Pictures Public Templates Videos
[emcleroy@rhel3 ~]$ cd Documents/
[emcleroy@rhel3 Documents]$ mkdir dns_server
[emcleroy@rhel3 Documents]$

Figure 5.41 – Creating the playbook directory

We are then going to create the inventory inside the new directory. This inventory example can be seen in the following screenshot:

```
[dns_server]
rhel1.example.com ansible_host=192.168.1.198
```

Figure 5.42 – Example of an inventory

First, we will start writing our playbook and then move on to creating the Jinja2 templates. As always, we will begin our playbook as follows:

```
---

- name: DNS server playbook
  hosts: rhel1.example.com
  become: true
  become_method: sudo
```

Next, we will move on to the tasks that we need to complete in order to install bind, which installs named.service, which runs the DNS server. We will then copy over the configuration files for named.conf and the zones. Then, we will update the firewall and test the name resolution. Please keep in mind that you will want to substitute any IP addresses for those of your systems and your DNS server for this playbook. The tasks are as follows:

```
tasks:
  - name: Install dns service
```

```
    package:
      name: bind
      state: latest


- name: Copy dns main config
  template:
    src: "{{ playbook_dir }}/named.conf.j2"
    dest: /etc/named.conf


- name: Copy dns forward zone configs
  template:
    src: "{{ playbook_dir }}/example.com.zone.j2"
    dest: /var/named/example.com.zone


- name: Copy dns reverse zone configs
  template:
    src: "{{ playbook_dir }}/1.168.192.zone.j2"
    dest: /var/named/1.168.192.zone


- name: Start and enable DNS
  service:
    name: named
    state: restarted
    enabled: true


- name: Enable DNS firewall rule
  command:
    cmd: firewall-cmd --permanent --add-service=dns


- name: Reload firewall
  command:
    cmd: firewall-cmd --reload


- name: Install DIG to check configuration
  package:
    name: bind-utils
```

```
        state: latest

    - name: Run dig to test configuration
      command:
        cmd: "dig rhel2.example.com @192.168.1.198"
```

Once that is complete, we will create our `named.conf` and zone files as `.j2` templates.

The `named.conf` file can be seen in the following screenshot:

```
//
// named.conf
//
// Provided by Red Hat bind package to configure the ISC BIND named(8) DNS
// server as a caching only nameserver (as a localhost DNS resolver only).
//
// See /usr/share/doc/bind*/sample/ for example named configuration files.
//

options {
        listen-on port 53 { any; };
        listen-on-v6 port 53 { any; };
        directory       "/var/named";
        dump-file       "/var/named/data/cache_dump.db";
        statistics-file "/var/named/data/named_stats.txt";
        memstatistics-file "/var/named/data/named_mem_stats.txt";
        secroots-file   "/var/named/data/named.secroots";
        recursing-file  "/var/named/data/named.recursing";
        allow-query     { localhost; 192.168.1.198; };

        recursion no;

        dnssec-enable yes;
        dnssec-validation yes;
-- INSERT --
```

Figure 5.43 – The named.conf file truncated for reading

The following screenshot shows the forward zone file:

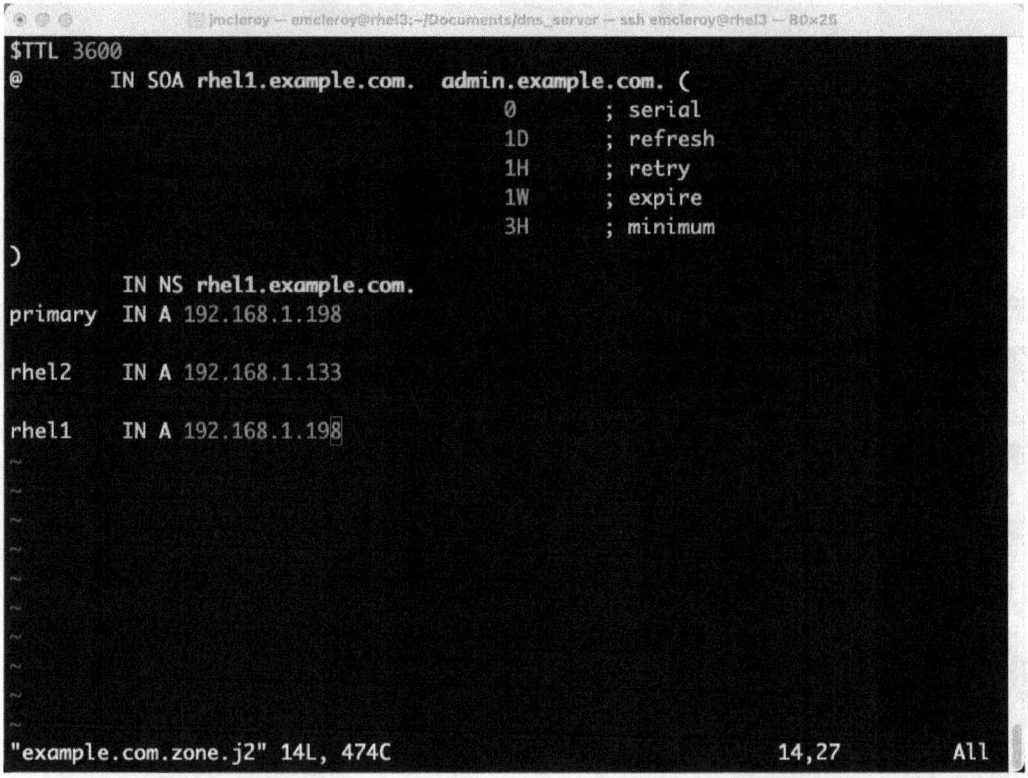

```
$TTL 3600
@        IN SOA rhel1.example.com.  admin.example.com. (
                                    0         ; serial
                                    1D        ; refresh
                                    1H        ; retry
                                    1W        ; expire
                                    3H        ; minimum
)
         IN NS rhel1.example.com.
primary  IN A 192.168.1.198

rhel2    IN A 192.168.1.133

rhel1    IN A 192.168.1.198
~
~
~
~
~
~
~
~
~
"example.com.zone.j2" 14L, 474C                        14,27        All
```

Figure 5.44 – Forward zone file example.com.zone.j2

Finally, we have the reverse zone file for the PTR records in the following screenshot:

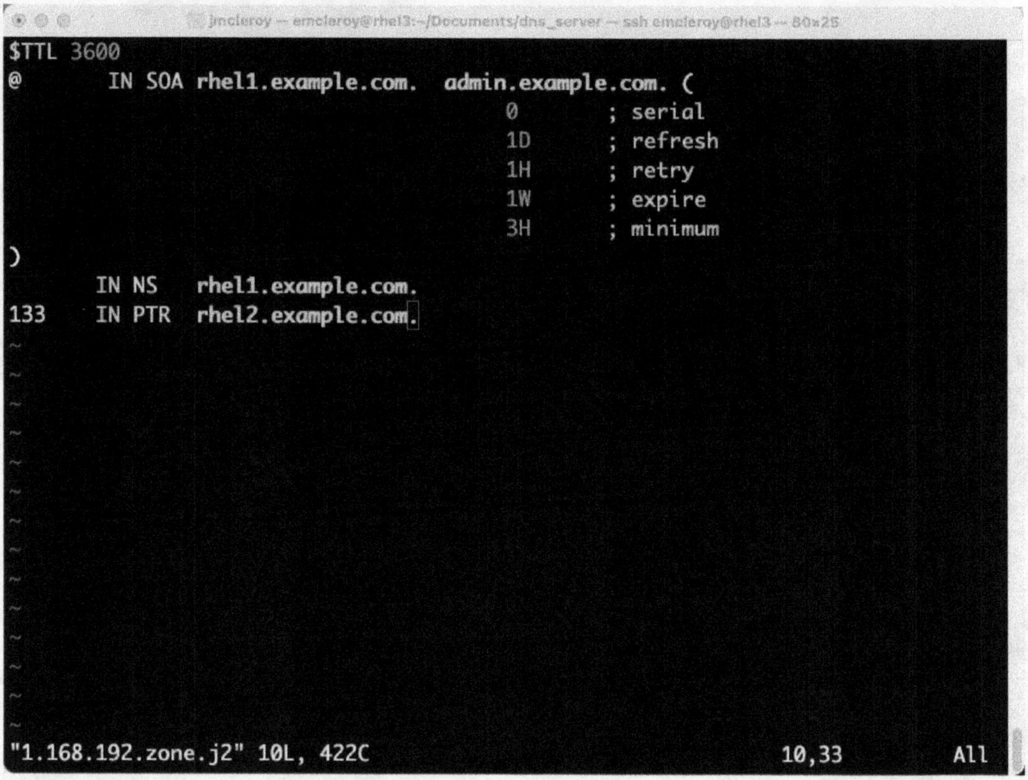

```
$TTL 3600
@        IN SOA rhel1.example.com.  admin.example.com. (
                                    0          ; serial
                                    1D         ; refresh
                                    1H         ; retry
                                    1W         ; expire
                                    3H         ; minimum
)
         IN NS    rhel1.example.com.
133      IN PTR   rhel2.example.com.

~
~
~
~
~
~
~
~
~
~
~
~
~
~
"1.168.192.zone.j2" 10L, 422C                              10,33         All
```

Figure 5.45 – Reverse zone file 1.168.192.zone.j2

After creating that, we will run the playbook using the following command again, which illustrates if you need to run the command without SSH keys set up. If you have SSH keys set up, you can omit -k -ask-become. Also, we are going to add verbosity so that we can see more of the standard out type of logs output by adding -v; this command is as follows:

```
$ ansible-playbook -i inventory dns_server_playbook.yml
--ask-become -u emcleroy -k -v
```

The following screenshot shows a successful playbook run:

```
jmcleroy — emcleroy@rhel3:-/Documents/dns_server — ssh emcleroy@rhel3 — 80×25
el2.example.com.\t\tIN\tA\n\n;; ANSWER SECTION:\nrhel2.example.com.\t3600\tIN\tA
\t192.168.1.133\n\n;; AUTHORITY SECTION:\nexample.com.\t\t3600\tIN\tNS\trhel1.ex
ample.com.\n\n;; ADDITIONAL SECTION:\nrhel1.example.com.\t3600\tIN\tA\t192.168.1
.198\n\n;; Query time: 0 msec\n;; SERVER: 192.168.1.198#53(192.168.1.198)\n;; WH
EN: Wed Aug 31 21:32:02 EDT 2022\n;; MSG SIZE  rcvd: 126", "stdout_lines": ["",
"; <<>> DiG 9.11.36-RedHat-9.11.36-3.el8 <<>> rhel2.example.com @192.168.1.198",
";; global options: +cmd", ";; Got answer:", ";; ->>HEADER<<- opcode: QUERY, st
atus: NOERROR, id: 47426", ";; flags: qr aa rd; QUERY: 1, ANSWER: 1, AUTHORITY:
1, ADDITIONAL: 2", ";; WARNING: recursion requested but not available", "", ";;
OPT PSEUDOSECTION:", "; EDNS: version: 0, flags:; udp: 1232", "; COOKIE: 91d14bc
8e4cce58465c0a31463100b92c2f51c475ca03100 (good)", ";; QUESTION SECTION:", ";rhe
l2.example.com.\t\tIN\tA", "", ";; ANSWER SECTION:", "rhel2.example.com.\t3600\t
IN\tA\t192.168.1.133", "", ";; AUTHORITY SECTION:", "example.com.\t\t3600\tIN\tN
S\trhel1.example.com.", "", ";; ADDITIONAL SECTION:", "rhel1.example.com.\t3600\
tIN\tA\t192.168.1.198", "", ";; Query time: 0 msec", ";; SERVER: 192.168.1.198#5
3(192.168.1.198)", ";; WHEN: Wed Aug 31 21:32:02 EDT 2022", ";; MSG SIZE  rcvd:
126"]}
META: ran handlers
META: ran handlers

PLAY RECAP ***********************************************************************
rhel1.example.com          : ok=10   changed=3   unreachable=0   failed=0   s
kipped=0   rescued=0   ignored=0

[emcleroy@rhel3 dns_server]$
```

Figure 5.46 – Playbook output

This wraps up working with Ansible Automation to automate your DNS servers. The use of Ansible Automation allows you to create these servers easily, and when you need to replicate them for any reason, such as **Disaster Recovery** (**DR**), you are able to do so quickly. This allows you to ensure that the configurations are copied over correctly every time without errors that can sometimes occur during human intervention.

Summary

In this chapter, we learned about IP addressing, DHCP, and DNS. We learned what is needed to get you started connecting on the internet and how it applies to you and your company. We were able to set up these programs to establish automatic DNS and DHCP via your very own servers. This allows you to control how your network is set up and how you connect to the intranet or internet. In the next chapter, we are going to learn about printers and email. The ability to set up these services is crucial to normal enterprise infrastructure and an important part of every company. Email servers and print servers are important as they allow the company to perform day-to-day business without faltering. I look forward to seeing you in the next chapter.

Part 3:
Red Hat Linux 8 – Configuring and Maintaining Applications with Automation and a Comprehensive Review with Exam Tips

In this part, you will learn to set up and maintain applications manually and automatically, and we will also provide a comprehensive review and tips and tricks to pass the EX358 exam.

This part contains the following chapters:

- *Chapter 6, Printer and Email – Setting up Printers and Email Services on Linux Servers*
- *Chapter 7, Databases – Setting up and Working with MariaDB SQL Databases*
- *Chapter 8, Web Servers and Web Traffic – Learning How to Create and Control Traffic*
- *Chapter 9, Comprehensive Review and Test Exam Questions*
- *Chapter 10, Tips and Tricks to Help with the Exam*

6

Printer and Email – Setting Up Printers and Email Services on Linux Servers

In this chapter, we are going to set up printer and email services. These services are vital to any company, big or small. The ability to print documentation is a necessity for many things. This can include documentation and handouts. Along with printing, we will be setting up email services for servers. This will allow mail servers to send out emails of the reports we create based on the type of reports we set up or programs that utilize email to complete tasks, such as alerting the user or providing configuration compliance reports. Through the use of email services, we will be able to ensure that nothing is overlooked when it comes to our systems.

In this chapter, we're going to cover the following main topics:

- Learning about printer services and setting them up manually

- Setting up printer services via Ansible Automation

- Learning about email services and setting them up manually

- Setting up email services via Ansible Automation

Technical requirements

You need to have applied the configuration from *Chapter 1, Block Storage – Learning How to Provision Block Storage on Red Hat Enterprise Linux*, in order to do the hands-on portion of this chapter. The relevant example files and playbooks can be found on the GitHub repository at `https://github.com/PacktPublishing/Red-Hat-Certified-Specialist-in-Services-Management-and-Automation-EX358-Exam-Guide/tree/main/Chapter06`.

Setting up GitHub access

Please refer to the instructions found in *Chapter 1, Block Storage – Learning How to Provision Block Storage on Red Hat Enterprise Linux*, to gain access to GitHub. You can find the Ansible automation playbooks for this chapter at `https://github.com/PacktPublishing/Red-Hat-Certified-Specialist-in-Services-Management-and-Automation-EX358-/tree/main/ch6`. Remember, these are only suggested playbooks, they do not represent the only way these solutions can be written.

You can always change them using raw, shell, or cmd to achieve the same results, but we seek to demonstrate the best way to accomplish our goals. Also keep in mind that we are not using the FCQN required in future versions of Ansible, as that will not be supported in the exam, which tests against Ansible 2.9.

Learning about printer services and setting them up manually

Systems need the ability to print in many cases. This could be in order to output some data that you need to deliver at a presentation, or due to the need to print material for later referencing. Through the use of printer services, we can set up and configure a print queue that will allow us to do just that. We will be able to control how and when the printer prints as well as which printer is set as the default. With this knowledge, you can set up printers on both RHEL desktops and servers. For the hands-on portion of the exercise, you will need a network printer on the same network as your servers.

Let's get started. First, we are going to install the `cups` package to provide the dependencies that we need to aggregate. We will do this with the commands in the following screenshot:

```
[emcleroy@rhel1 ~]$ sudo dnf install cups
Updating Subscription Management repositories.
Last metadata expiration check: 0:02:49 ago on Wed 31 Aug 2022 11:00:34 PM EDT.
Dependencies resolved.
================================================================================
 Package          Arch     Version             Repository                  Size
================================================================================
Installing:
 cups             x86_64   1:2.2.6-45.el8_6.2  rhel-8-for-x86_64-appstream-rpms 1.4 M
Installing dependencies:
 qpdf-libs        x86_64   7.1.1-10.el8        rhel-8-for-x86_64-appstream-rpms 338 k
 ghostscript      x86_64   9.25-5.el8          rhel-8-for-x86_64-appstream-rpms  82 k
 poppler-utils    x86_64   0.66.0-26.el8       rhel-8-for-x86_64-appstream-rpms 228 k
 cups-filters-libs
                  x86_64   1.20.0-20.el8       rhel-8-for-x86_64-appstream-rpms 135 k
 cups-filters     x86_64   1.20.0-20.el8       rhel-8-for-x86_64-appstream-rpms 779 k
 cups-client      x86_64   1:2.2.6-45.el8_6.2  rhel-8-for-x86_64-appstream-rpms 170 k

Transaction Summary
================================================================================
Install  7 Packages

Total download size: 3.1 M
Installed size: 12 M
Is this ok [y/N]:
```

Figure 6.1 – Install cups package

By installing the cups package, we install all the packages needed to create print queues and query printers directly attached to other servers or print servers. We will need to enable and start the service, as shown in the following screenshot:

```
  ● ◯ ◯              jmcleroy — emcleroy@rhel1:~ — ssh emcleroy@rhel1 — 80×25
  Installing       : qpdf-libs-7.1.1-10.el8.x86_64                        5/7
  Running scriptlet: qpdf-libs-7.1.1-10.el8.x86_64                        5/7
  Installing       : cups-1:2.2.6-45.el8_6.2.x86_64                       6/7
  Running scriptlet: cups-1:2.2.6-45.el8_6.2.x86_64                       6/7
  Installing       : cups-filters-1.20.0-20.el8.x86_64                    7/7
  Running scriptlet: cups-filters-1.20.0-20.el8.x86_64                    7/7
  Verifying        : qpdf-libs-7.1.1-10.el8.x86_64                        1/7
  Verifying        : ghostscript-9.25-5.el8.x86_64                        2/7
  Verifying        : poppler-utils-0.66.0-26.el8.x86_64                   3/7
  Verifying        : cups-filters-libs-1.20.0-20.el8.x86_64               4/7
  Verifying        : cups-filters-1.20.0-20.el8.x86_64                    5/7
  Verifying        : cups-client-1:2.2.6-45.el8_6.2.x86_64                6/7
  Verifying        : cups-1:2.2.6-45.el8_6.2.x86_64                       7/7
Installed products updated.

Installed:
  cups-1:2.2.6-45.el8_6.2.x86_64            qpdf-libs-7.1.1-10.el8.x86_64
  ghostscript-9.25-5.el8.x86_64            poppler-utils-0.66.0-26.el8.x86_64
  cups-filters-libs-1.20.0-20.el8.x86_64   cups-filters-1.20.0-20.el8.x86_64
  cups-client-1:2.2.6-45.el8_6.2.x86_64

Complete!
[emcleroy@rhel1 ~]$ sudo systemctl enable cups
[emcleroy@rhel1 ~]$ sudo systemctl start cups
[emcleroy@rhel1 ~]$ ▯
```

Figure 6.2 – Start and enable cups for printer setup

Next, we are going to enable firewall rules to allow 631/tcp, which is the typical port that printers serve from. We also enable mDNS, which allows dynamic discovery through the **Internet Printing Protocol** (**IPP**). The commands needed are in the following screenshot:

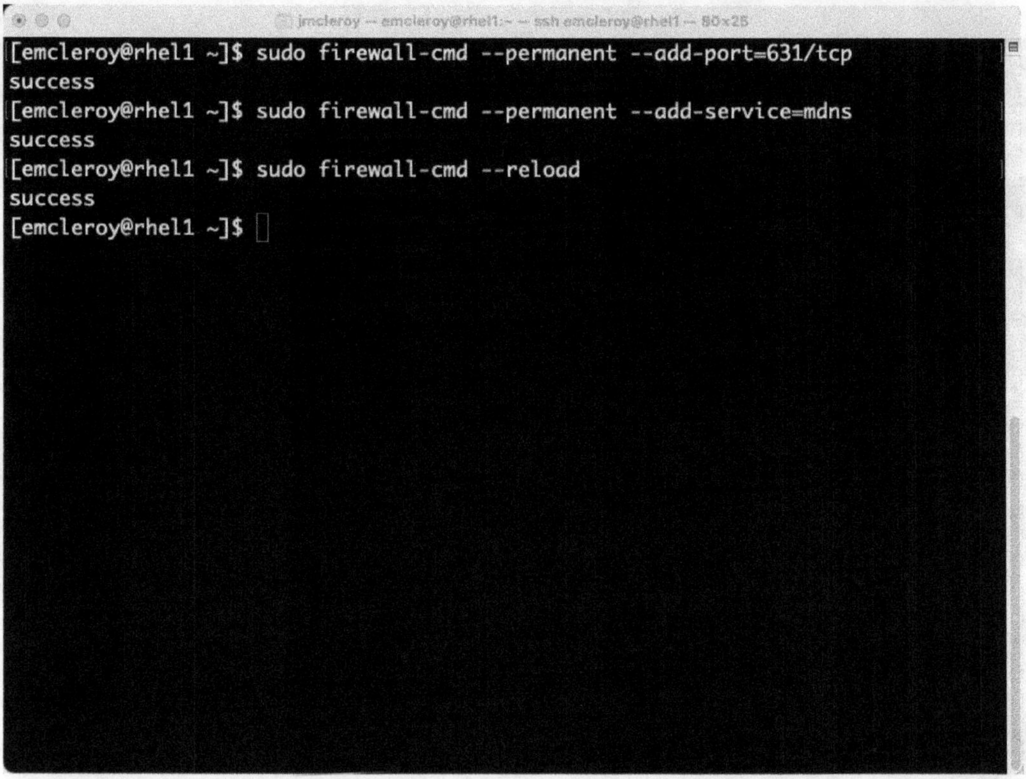

```
[emcleroy@rhel1 ~]$ sudo firewall-cmd --permanent --add-port=631/tcp
success
[emcleroy@rhel1 ~]$ sudo firewall-cmd --permanent --add-service=mdns
success
[emcleroy@rhel1 ~]$ sudo firewall-cmd --reload
success
[emcleroy@rhel1 ~]$ 
```

Figure 6.3 – Add the required firewall rules to discover printers and printing

The next step for setting up printing on your system is to look up network printers. We will use the ippfind tool to do that with the command shown in the following screenshot:

Figure 6.4 – Use the ippfind tool to look up network printers

After we have located our printers, we will take the one we would like to target and create a print queue for it. This will allow us to queue up print jobs to send to the printer. This provides the ability to enable and disable the printer's printing functionality, while maintaining the print queue when the printer is offline or disabled (unless you intervene manually or through Ansible Automation). We can see the print queue created with the lpadmin command. (Note that is an *L* at the start of lpadmin.) We will utilize the following flags: -p, -v, -m, and -E. These mean the following: -p queue name, -v device url, while -m everywhere is the IPP definition, and -E is for immediate enabling of the printer queue. Please first ensure that the DNS name is routable for the printer or add it to /etc/hosts in order to ensure reachability. The full command utilized can be seen in the following screenshot:

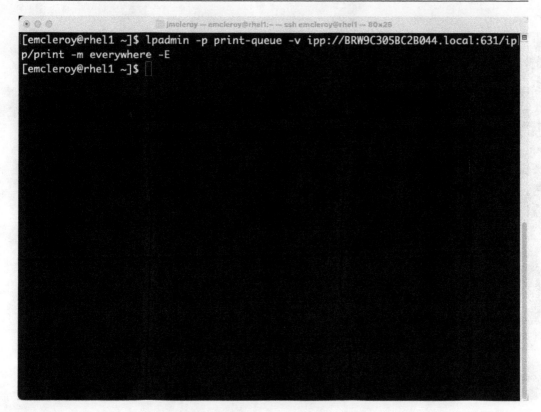

Figure 6.5 – Configure the print queue

Next, we are going to set the default printer so that when you create a print job it is sent to the right printer. This is normally used when you have multiple printers and need to ensure you print to the right one, perhaps at a specific location in a building or due to needing to utilize a certain type of printer. The command to set the default printer can be seen in the following screenshot:

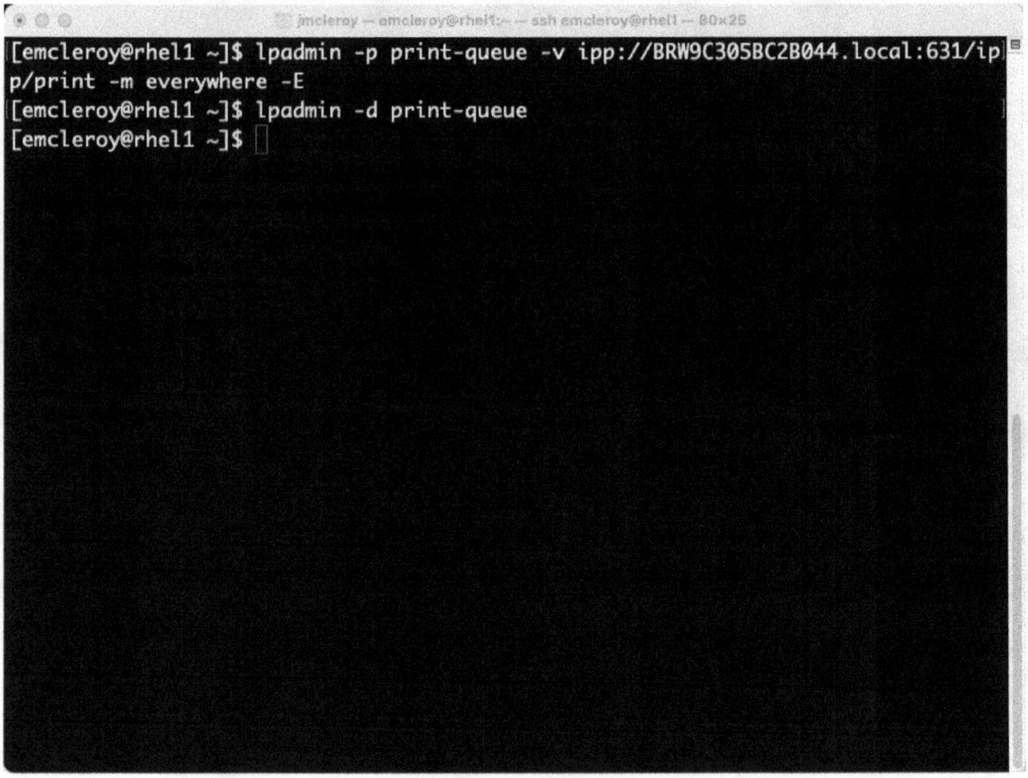

```
[emcleroy@rhel1 ~]$ lpadmin -p print-queue -v ipp://BRW9C305BC2B044.local:631/ip
p/print -m everywhere -E
[emcleroy@rhel1 ~]$ lpadmin -d print-queue
[emcleroy@rhel1 ~]$ ▯
```

Figure 6.6 – Set the default printer

Next, we are going to create a simple text file to print and test that our print queue is set up correctly. In my case, my file looks like the following screenshot:

Figure 6.7 – Create a test document to print

After we have created the text file to print, we are going to go and actually print it. This is done with the command found in the following screenshot:

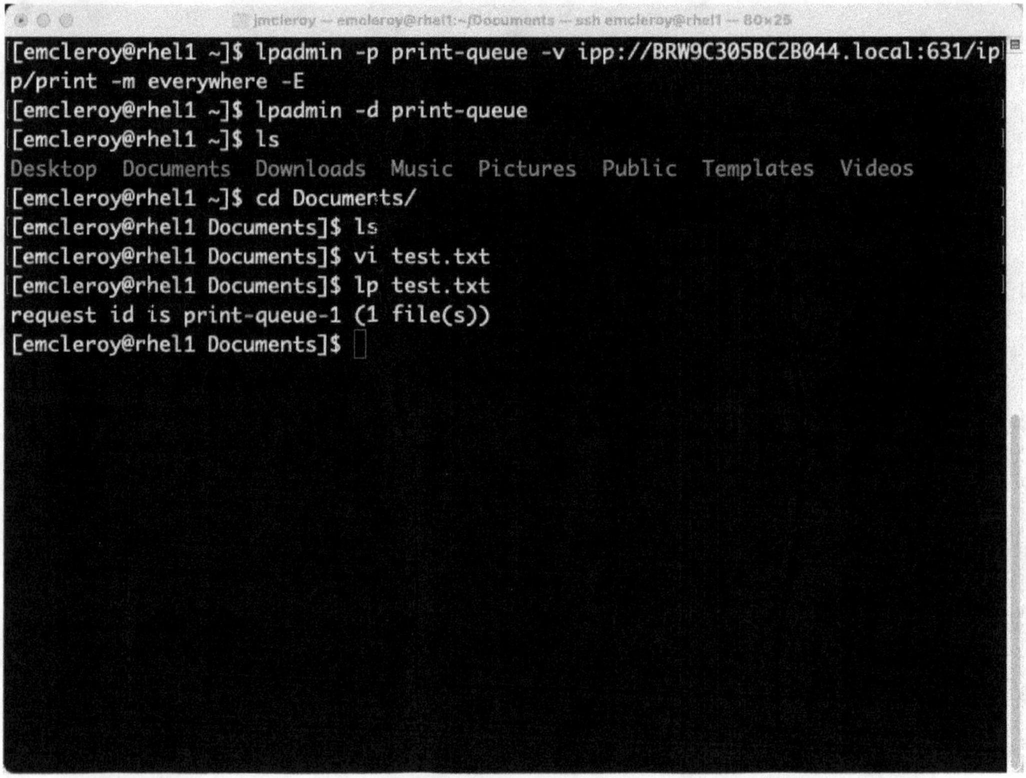
jmcleroy — emcleroy@rhel1:~/Documents — ssh emcleroy@rhel1 — 80×25
[emcleroy@rhel1 ~]$ lpadmin -p print-queue -v ipp://BRW9C305BC2B044.local:631/ip
p/print -m everywhere -E
[emcleroy@rhel1 ~]$ lpadmin -d print-queue
[emcleroy@rhel1 ~]$ ls
Desktop Documents Downloads Music Pictures Public Templates Videos
[emcleroy@rhel1 ~]$ cd Documents/
[emcleroy@rhel1 Documents]$ ls
[emcleroy@rhel1 Documents]$ vi test.txt
[emcleroy@rhel1 Documents]$ lp test.txt
request id is print-queue-1 (1 file(s))
[emcleroy@rhel1 Documents]$

Figure 6.8 – Print test document

If the printer successfully printed your page, then you are able to create and control printer jobs on `rhel1.example.com`.

Next, let's see how to disable and enable the print queue. When you disable the printer, it will queue the print jobs instead of sending them to the printer. This can be seen in the following screenshot:

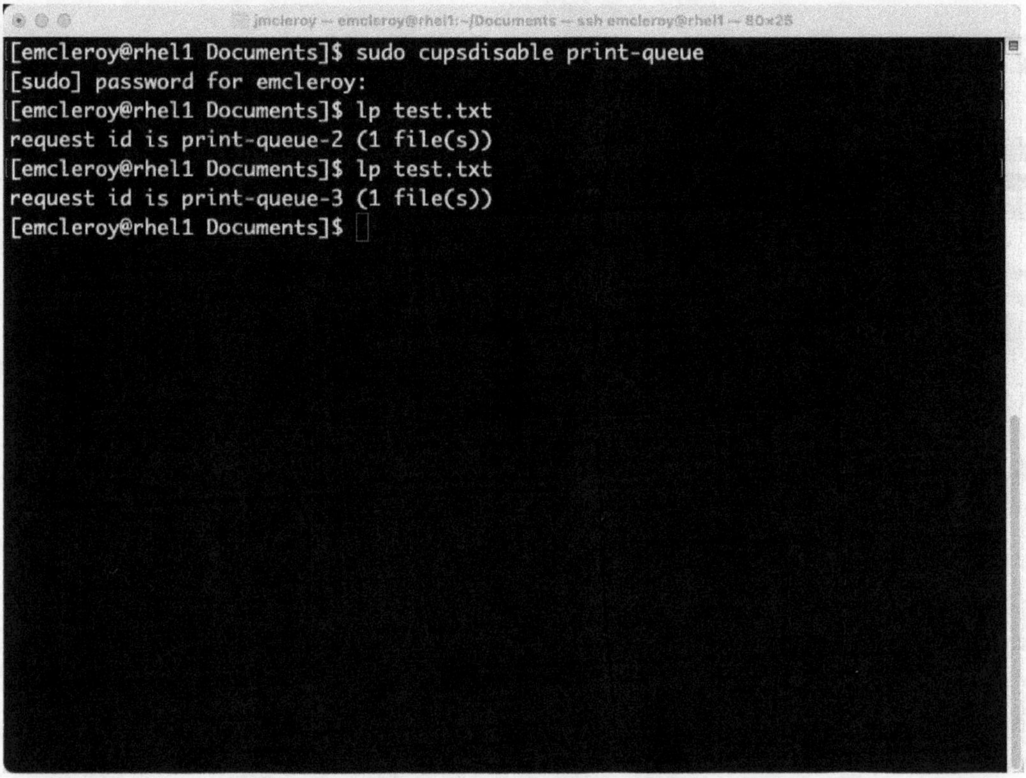
jmcleroy — emcleroy@rhel1:~/Documents — ssh emcleroy@rhel1 — 80×25
[emcleroy@rhel1 Documents]$ sudo cupsdisable print-queue
[sudo] password for emcleroy:
[emcleroy@rhel1 Documents]$ lp test.txt
request id is print-queue-2 (1 file(s))
[emcleroy@rhel1 Documents]$ lp test.txt
request id is print-queue-3 (1 file(s))
[emcleroy@rhel1 Documents]$

Figure 6.9 – Disable printer and then attempt to print

We are then going to check the current print queue and how it is not moving forward as the printer is disabled. After that, we will cancel the print jobs so that they do not print when we re-enable the printer. Going to the home directory and using the commands found in the following screenshot will show the print queue and cancel the print jobs:

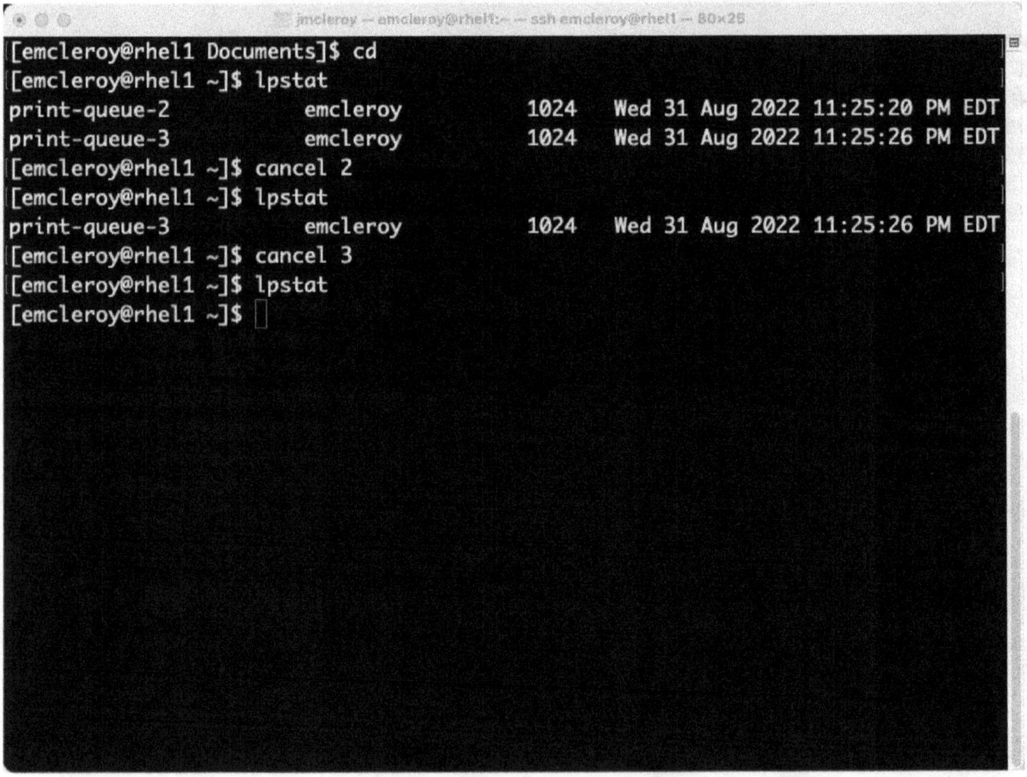

```
[emcleroy@rhel1 Documents]$ cd
[emcleroy@rhel1 ~]$ lpstat
print-queue-2              emcleroy        1024   Wed 31 Aug 2022 11:25:20 PM EDT
print-queue-3              emcleroy        1024   Wed 31 Aug 2022 11:25:26 PM EDT
[emcleroy@rhel1 ~]$ cancel 2
[emcleroy@rhel1 ~]$ lpstat
print-queue-3              emcleroy        1024   Wed 31 Aug 2022 11:25:26 PM EDT
[emcleroy@rhel1 ~]$ cancel 3
[emcleroy@rhel1 ~]$ lpstat
[emcleroy@rhel1 ~]$
```

Figure 6.10 – Print queue shows stalled jobs due to the printer being disabled

To ensure that we will still be able to print, next we are going to re-enable the print queue. This will allow any print jobs to be sent to our default printer. This is accomplished by the command in the following screenshot:

```
[emcleroy@rhel1 Documents]$ cd
[emcleroy@rhel1 ~]$ lpstat
print-queue-2              emcleroy        1024    Wed 31 Aug 2022 11:25:20 PM EDT
print-queue-3              emcleroy        1024    Wed 31 Aug 2022 11:25:26 PM EDT
[emcleroy@rhel1 ~]$ cancel 2
[emcleroy@rhel1 ~]$ lpstat
print-queue-3              emcleroy        1024    Wed 31 Aug 2022 11:25:26 PM EDT
[emcleroy@rhel1 ~]$ cancel 3
[emcleroy@rhel1 ~]$ lpstat
[emcleroy@rhel1 ~]$ sudo cupsenable print-queue
[emcleroy@rhel1 ~]$ 
```

Figure 6.11 – Re-enable the print queue after clearing the print queue

In this section, we set up a print queue to allow a Linux device to print to a network printer. This information is crucial when you need to be able to create physical documentation. This could be for handouts at a company event or for an internal meeting. The ability to print is necessary and knowing how to set up a print queue enables that ability. Next, we are going to set up a print queue using Ansible Automation.

Setting up printer services via Ansible Automation

We are going to start this section off by creating a playbook for enabling print services using `cups`. We will then create playbooks to enable and disable the print queues and also to drain any queued print jobs that might be in line to print. This will allow us to control setting up printers over many devices via automation, which will shorten the time to completion considerably.

First, we will start with a playbook directory, inside of which we will have an inventory file with a list of the servers we want to set up services on. In our case, we are going to set up a print queue on `rhel1.example.com` and `rhel2.example.com`. This will allow us to showcase using Ansible

Automation in more than just a single playbook. We are going to create a `cups_playbook` directory, inside of which we are going to first create the inventory file shown in the following screenshot:

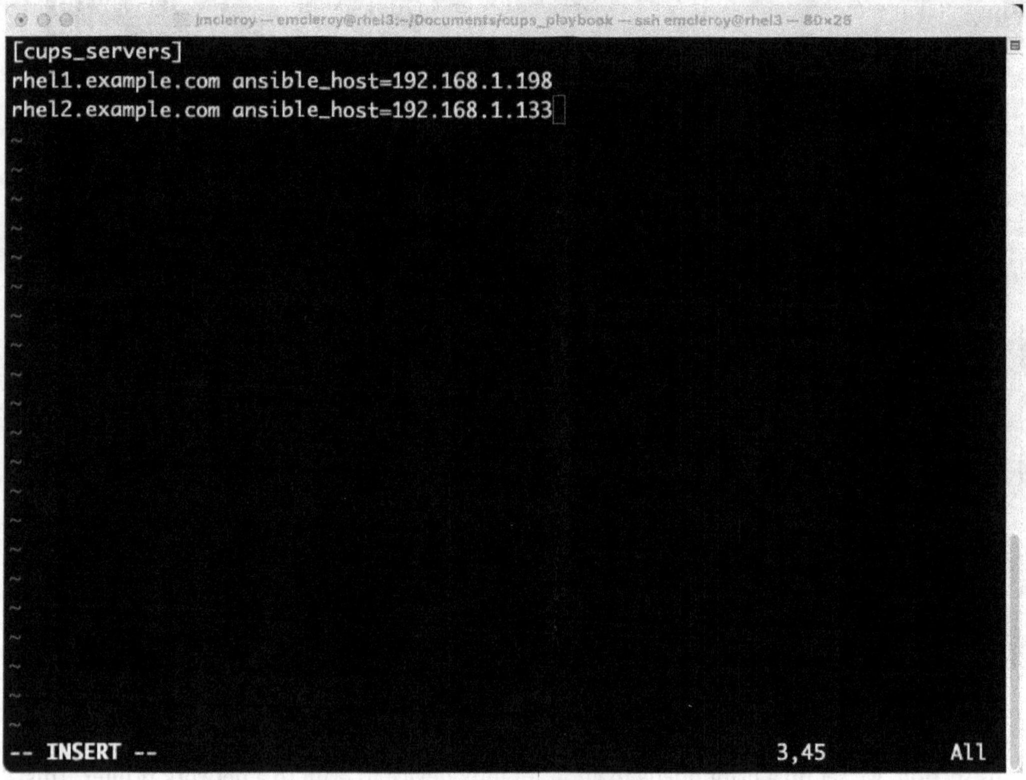

Figure 6.12 – Inventory file for playbook

Next, we are going to start the playbook as we always have by creating the top level of the playbook. The following shows the format that is required for the beginning of all playbooks when using Ansible Automation. The start of the playbook can be seen in the following code:

```
---

- name: Cups printer queue setup
  hosts: cups_servers
  become: true
  become_method: sudo
```

Next, let's move on to the tasks. These include installing cups, configuring firewall rules, and setting up the printer queue. For this, you will need to know the URL of the printer ahead of time, so you might need to gather that information from the server using the following code:

```
[emcleroy@rhel1 ~]$ ippfind -T 20
ipp://HPE070EA601518.local:631/ipp/print
ipp://EPSON04046D.local:631/ipp/print
ipp://BRW9C305BC2B044.local:631/ipp/print
```

In my case, I will be using the Brother printer that is at the bottom of the printer-finding command output in the preceding snippet.

With that information, we can write out the command that we learned earlier to set up the print queue. You can see the tasks for installing cups, firewall configuration, and print queue setup in the following code:

```
  tasks:
    - name: Install cups
      package:
        name: cups
        state: latest

    - name: Start and enable cups service
      service:
        name: cups
        state: started
        enabled: true

    - name: Enable firewall rules
      firewalld:
        service: mdns
        permanent: true
        state: enabled

    - name: Enable firewall rules
      firewalld:
        port: 631/tcp
        permanent: true
```

```
         state: enabled

  - name: Reload firewall
    command:
      cmd: firewall-cmd --reload

  - name: Setup print queue
    command:
      cmd: lpadmin -p printer-queue -v ipp://BRW9C305BC2B044.
local:631/ipp/print -m everywhere -E

  - name: Set default print queue
    command:
      cmd: lpadmin -d printer-queue

  - name: Test to ensure print queue is setup correctly
    command:
      cmd: lpstat -p printer-queue
```

We will run the playbook with the following command, including the -v flag for verbose output such as when the print queue is shown enabled:

```
[emcleroy@rhel3 cups_playbook]$ ansible-playbook -i inventory
cups_playbook.yml -u emcleroy -k --ask-become -v
```

The output from this successful playbook run is truncated but you can see where the print queue was enabled in the following screenshot:

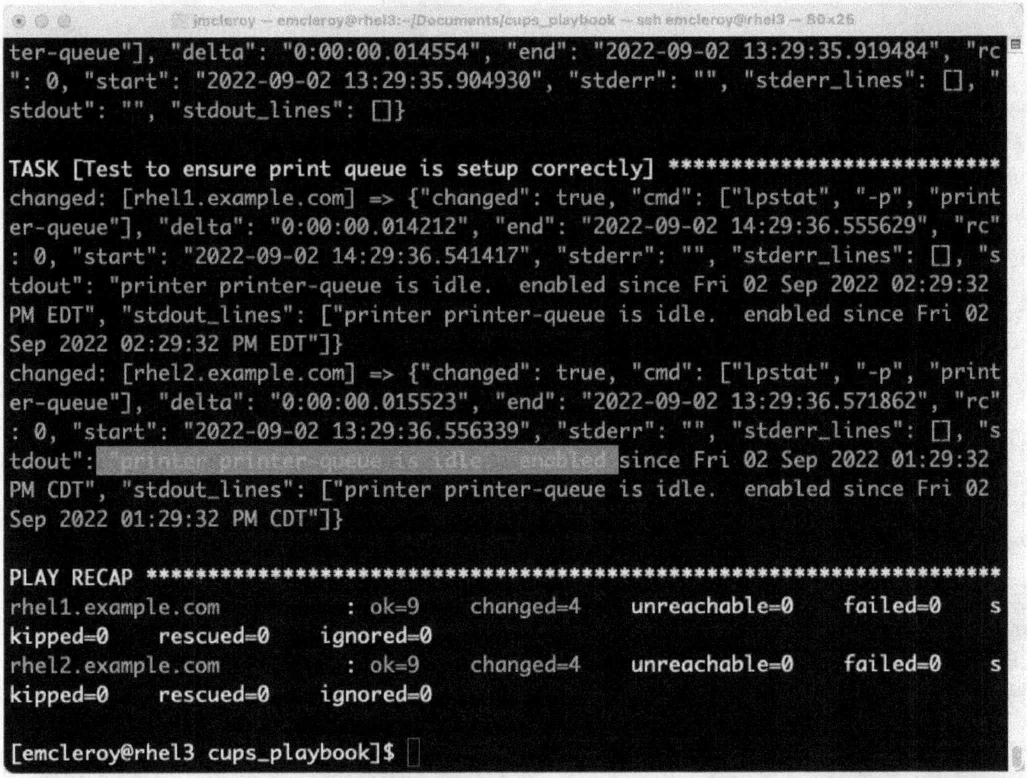
jmcleroy — emcleroy@rhel3:~/Documents/cups_playbook — ssh emcleroy@rhel3 — 80x25
ter-queue"], "delta": "0:00:00.014554", "end": "2022-09-02 13:29:35.919484", "rc
": 0, "start": "2022-09-02 13:29:35.904930", "stderr": "", "stderr_lines": [], "
stdout": "", "stdout_lines": []}
TASK [Test to ensure print queue is setup correctly] ***************************
changed: [rhel1.example.com] => {"changed": true, "cmd": ["lpstat", "-p", "print
er-queue"], "delta": "0:00:00.014212", "end": "2022-09-02 14:29:36.555629", "rc"
: 0, "start": "2022-09-02 14:29:36.541417", "stderr": "", "stderr_lines": [], "s
tdout": "printer printer-queue is idle. enabled since Fri 02 Sep 2022 02:29:32
PM EDT", "stdout_lines": ["printer printer-queue is idle. enabled since Fri 02
Sep 2022 02:29:32 PM EDT"]}
changed: [rhel2.example.com] => {"changed": true, "cmd": ["lpstat", "-p", "print
er-queue"], "delta": "0:00:00.015523", "end": "2022-09-02 13:29:36.571862", "rc"
: 0, "start": "2022-09-02 13:29:36.556339", "stderr": "", "stderr_lines": [], "s
tdout": "printer printer-queue is idle. enabled since Fri 02 Sep 2022 01:29:32
PM CDT", "stdout_lines": ["printer printer-queue is idle. enabled since Fri 02
Sep 2022 01:29:32 PM CDT"]}
PLAY RECAP ***
rhel1.example.com : ok=9 changed=4 unreachable=0 failed=0 s
kipped=0 rescued=0 ignored=0
rhel2.example.com : ok=9 changed=4 unreachable=0 failed=0 s
kipped=0 rescued=0 ignored=0
[emcleroy@rhel3 cups_playbook]$

Figure 6.13 – Successful playbook run for the cups create playbook

Next, we are going to write a playbook to disable the print queue and then another to enable it. Here you can see the playbook code to disable the print queue:

```
---
- name: Cups printer queue setup
  hosts: cups_servers
  become: true
  become_method: sudo

  tasks:
    - name: Disable print queue
      command:
        cmd: cupsdisable printer-queue
```

The output of this playbook run for disabling the print queue can be seen in the following screenshot:

```
⊙ ○ ○          |mcleroy — emcleroy@rhel3:~/Documents/cups_playbook — ssh emcleroy@rhel2 — 80×25

PLAY [Cups printer queue setup] ****************************************************

TASK [Gathering Facts] *************************************************************
ok: [rhel2.example.com]
ok: [rhel1.example.com]

TASK [Disable print queue] *********************************************************
changed: [rhel1.example.com] => {"changed": true, "cmd": ["cupsdisable", "printe
r-queue"], "delta": "0:00:00.014244", "end": "2022-09-02 14:36:09.302002", "rc":
 0, "start": "2022-09-02 14:36:09.287758", "stderr": "", "stderr_lines": [], "st
dout": "", "stdout_lines": []}
changed: [rhel2.example.com] => {"changed": true, "cmd": ["cupsdisable", "printe
r-queue"], "delta": "0:00:00.023546", "end": "2022-09-02 13:36:09.323062", "rc":
 0, "start": "2022-09-02 13:36:09.299516", "stderr": "", "stderr_lines": [], "st
dout": "", "stdout_lines": []}

PLAY RECAP *************************************************************************
rhel1.example.com          : ok=2    changed=1    unreachable=0    failed=0    s
kipped=0    rescued=0    ignored=0
rhel2.example.com          : ok=2    changed=1    unreachable=0    failed=0    s
kipped=0    rescued=0    ignored=0

[emcleroy@rhel3 cups_playbook]$ ansible-playbook -i inventory cups_playbook.yml
-u emcleroy -k --ask-become -v
```

Figure 6.14 – Successful playbook run for the cups disable playbook

The playbook shown in the following code enables the print queue:

```
---

- name: Cups printer queue setup
  hosts: cups_servers
  become: true
  become_method: sudo

  tasks:
    - name: Enable print queue
      command:
        cmd: cupsenable printer-queue
```

The output for enabling the print queue can be seen in the following screenshot:

```
jmcleroy — emcleroy@rhel3:~/Documents/cups_playbook — ssh emcleroy@rhel3 — 80×25

BECOME password[defaults to SSH password]:

PLAY [Cups printer queue setup] ********************************************

TASK [Gathering Facts] *****************************************************
ok: [rhel1.example.com]
ok: [rhel2.example.com]

TASK [Disable print queue] *************************************************
changed: [rhel1.example.com] => {"changed": true, "cmd": ["cupsdisable", "printe
r-queue"], "delta": "0:00:00.024488", "end": "2022-09-02 14:43:19.614771", "rc":
 0, "start": "2022-09-02 14:43:19.590283", "stderr": "", "stderr_lines": [], "st
dout": "", "stdout_lines": []}
changed: [rhel2.example.com] => {"changed": true, "cmd": ["cupsdisable", "printe
r-queue"], "delta": "0:00:00.021162", "end": "2022-09-02 13:43:19.652202", "rc":
 0, "start": "2022-09-02 13:43:19.631040", "stderr": "", "stderr_lines": [], "st
dout": "", "stdout_lines": []}

PLAY RECAP *****************************************************************
rhel1.example.com          : ok=2    changed=1    unreachable=0    failed=0    s
kipped=0    rescued=0    ignored=0
rhel2.example.com          : ok=2    changed=1    unreachable=0    failed=0    s
kipped=0    rescued=0    ignored=0

[emcleroy@rhel3 cups_playbook]$
```

Figure 6.15 – Successful playbook run for the cups enable playbook

In this section, we learned how to set up print queues through Ansible Automation. This can be extremely useful in an enterprise environment where you might be setting up many servers that need to print to the same printer. By setting up Ansible Automation playbooks to do the work for you, you free up your time for other more meaningful work.

Learning about email services and setting them up manually

Email services are an essential part of any infrastructure. The ability to send email reports or alerts to a dedicated email list is one of the reasons that email is so useful. Using the postfix package, you are able to set up full email relays or null clients that forward emails to other relays. This allows you to control how your email is routed in your network. In this section, we will see how to set up a null client for these purposes. We will install postfix to control email from the server, set up firewall rules, and finally set up the null client settings both manually and through Ansible Automation.

The first thing we have to do is install postfix. Running the command to install postfix can be seen in the following screenshot:

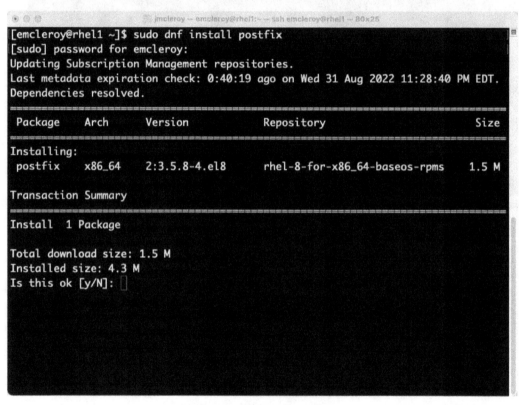

Figure 6.16 – Install postfix

We are then going to view the configuration file for postfix. This is not the best way to set up postfix as the postconf command is normally used to change the required settings. The configuration file can be seen truncated in the following screenshot and is located at /etc/postfix/main.cf:

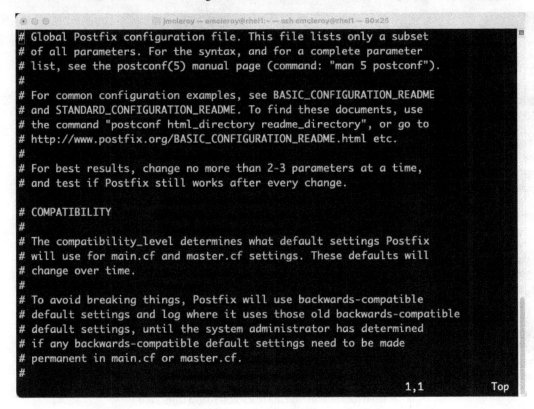

Figure 6.17 – Postfix main.cf truncated

Using the `postconf` command, you can see the same configuration without the comments, as shown in the following screenshot:

```
[emcleroy@rhel1 ~]$ postconf
2bounce_notice_recipient = postmaster
access_map_defer_code = 450
access_map_reject_code = 554
address_verify_cache_cleanup_interval = 12h
address_verify_default_transport = $default_transport
address_verify_local_transport = $local_transport
address_verify_map = btree:$data_directory/verify_cache
address_verify_negative_cache = yes
address_verify_negative_expire_time = 3d
address_verify_negative_refresh_time = 3h
address_verify_pending_request_limit = 5000
address_verify_poll_count = ${stress?{1}:{3}}
address_verify_poll_delay = 3s
address_verify_positive_expire_time = 31d
address_verify_positive_refresh_time = 7d
address_verify_relay_transport = $relay_transport
address_verify_relayhost = $relayhost
address_verify_sender = $double_bounce_sender
address_verify_sender_dependent_default_transport_maps = $sender_dependent_defau
lt_transport_maps
address_verify_sender_dependent_relayhost_maps = $sender_dependent_relayhost_map
s
address_verify_sender_ttl = 0s
address_verify_service_name = verify
```

Figure 6.18 – Output from the postconf command truncated

Next, we are going to make the required changes to the system to set up what is needed for `postfix` to work as a null client, as shown in the following screenshot:

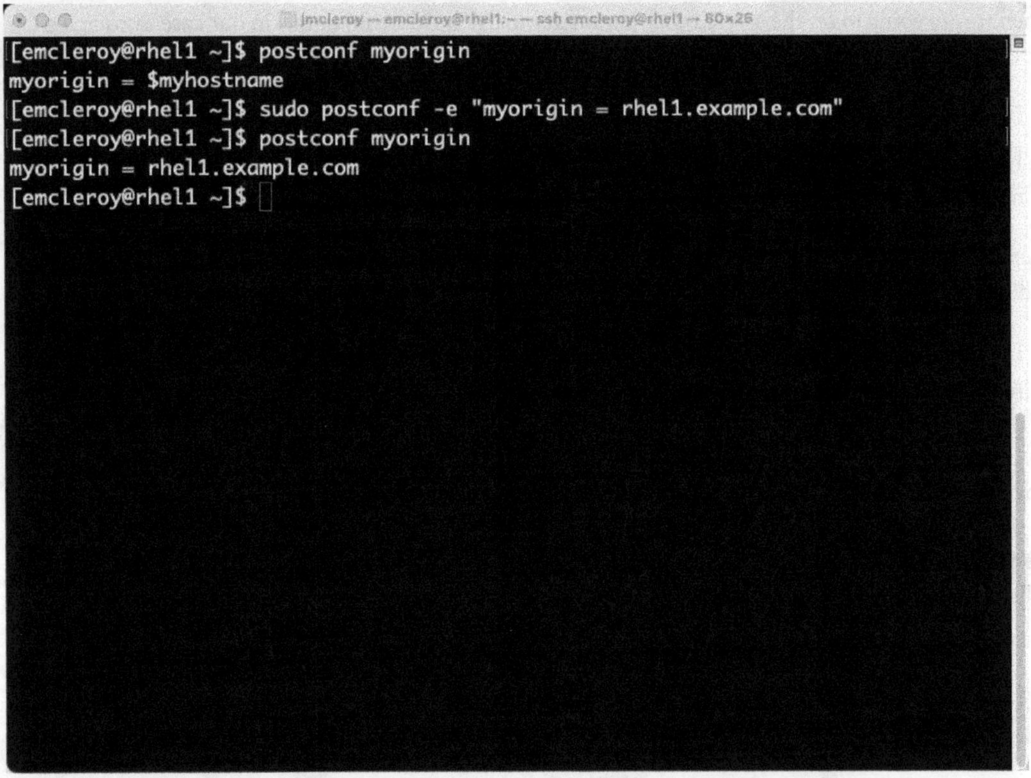

```
[emcleroy@rhel1 ~]$ postconf myorigin
myorigin = $myhostname
[emcleroy@rhel1 ~]$ sudo postconf -e "myorigin = rhel1.example.com"
[emcleroy@rhel1 ~]$ postconf myorigin
myorigin = rhel1.example.com
[emcleroy@rhel1 ~]$ 
```

Figure 6.19 – Using the postconf command to change configuration values

This allows you to make changes to your postfix configuration based on the requirements set by Red Hat during your exam.

Let's look at an example of what you may need to set for a null client setup. A null client is just a server that passes emails on to another server and does not accept any email itself.

The first setting to change is inet_interfaces, making it loopback-only, as shown in the following command:

```
[emcleroy@rhel1 ~]$ sudo postconf -e "inet_interfaces =
loopback-only"
[sudo] password for emcleroy:
[emcleroy@rhel1 ~]$ sudo postconf inet_interfaces
inet_interfaces = loopback-only
```

Then we are going to set the `myorigin` parameter to use the domain name, in our case, `example.com`, as seen in the following code:

```
[emcleroy@rhel1 ~]$ sudo postconf -e "myorigin = example.com"
[emcleroy@rhel1 ~]$ sudo postconf myorigin
myorigin = example.com
```

Next, we are going to set `inet_protocols` to `ipv4` purely for simplicity, as shown in the following code:

```
[emcleroy@rhel1 ~]$ sudo postconf -e "inet_protocols = ipv4"
[emcleroy@rhel1 ~]$ sudo postconf inet_protocols
inet_protocols = ipv4
```

The next setting we are going to change is the `mydestination` parameter, as `postfix` is allowed to deliver mail if this value is populated. To ensure that nothing is delivered to this server locally, we will remove all settings. We are going to change the `mydestination` setting to blank, per the following code:

```
[emcleroy@rhel1 ~]$ sudo postconf -e "mydestination ="
[emcleroy@rhel1 ~]$ sudo postconf mydestination
mydestination =
```

Next, let's alter the `mynetworks` parameter in order to submit messages to the mail relay for any network hosts. We will set this to our loopback IP in order for it to take effect, as seen in the following code:

```
[emcleroy@rhel1 ~]$ sudo postconf -e "mynetworks = 127.0.0.0/8"
[emcleroy@rhel1 ~]$ sudo postconf mynetworks
mynetworks = 127.0.0.0/8
```

The next thing we are going to set is an error message for local delivery. This helps us confirm that local delivery is disabled. This is accomplished in the following code:

```
[emcleroy@rhel1 ~]$ sudo postconf -e "local_transport = error:
no local delivery"
[emcleroy@rhel1 ~]$ sudo postconf local_transport
local_transport = error: no local delivery
```

Finally, we are going to point to an active mail relay that can deliver email locally and to the company's mail servers. This is done with the following code:

```
[emcleroy@rhel1 ~]$ sudo postconf -e "relayhost = [rhel2.
example.com]"
[emcleroy@rhel1 ~]$ sudo postconf relayhost
relayhost = [rhel2.example.com]
```

We have now set up the postfix mail service. Now, let's open the firewall, start the service, and enable it. These commands can be seen in the following screenshot:

Figure 6.20 – Enable and start the service, add firewall rules, and reload the firewall

If you start and enable the postfix service before configuring, you need to make sure to reload the service again to ensure that the changes have been accepted and check the status, as shown in the following screenshot:

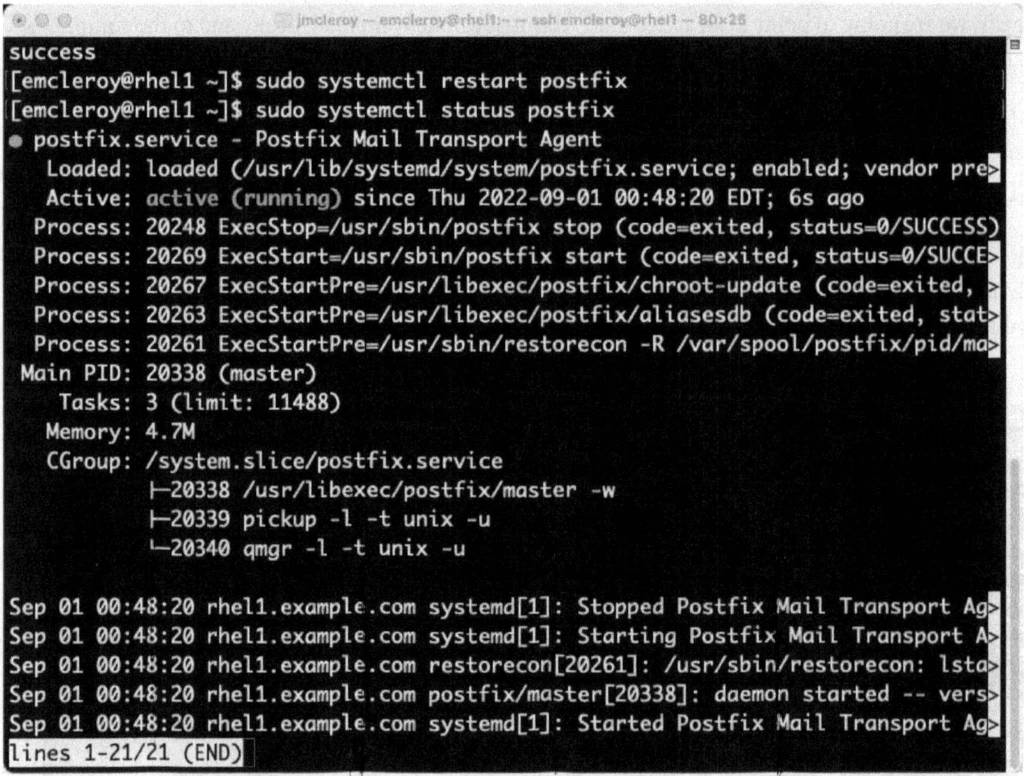

```
success
[emcleroy@rhel1 ~]$ sudo systemctl restart postfix
[emcleroy@rhel1 ~]$ sudo systemctl status postfix
● postfix.service - Postfix Mail Transport Agent
   Loaded: loaded (/usr/lib/systemd/system/postfix.service; enabled; vendor pre>
   Active: active (running) since Thu 2022-09-01 00:48:20 EDT; 6s ago
  Process: 20248 ExecStop=/usr/sbin/postfix stop (code=exited, status=0/SUCCESS)
  Process: 20269 ExecStart=/usr/sbin/postfix start (code=exited, status=0/SUCCE>
  Process: 20267 ExecStartPre=/usr/libexec/postfix/chroot-update (code=exited, >
  Process: 20263 ExecStartPre=/usr/libexec/postfix/aliasesdb (code=exited, stat>
  Process: 20261 ExecStartPre=/usr/sbin/restorecon -R /var/spool/postfix/pid/ma>
 Main PID: 20338 (master)
    Tasks: 3 (limit: 11488)
   Memory: 4.7M
   CGroup: /system.slice/postfix.service
           ├─20338 /usr/libexec/postfix/master -w
           ├─20339 pickup -l -t unix -u
           └─20340 qmgr -l -t unix -u

Sep 01 00:48:20 rhel1.example.com systemd[1]: Stopped Postfix Mail Transport Ag>
Sep 01 00:48:20 rhel1.example.com systemd[1]: Starting Postfix Mail Transport A>
Sep 01 00:48:20 rhel1.example.com restorecon[20261]: /usr/sbin/restorecon: lsta>
Sep 01 00:48:20 rhel1.example.com postfix/master[20338]: daemon started -- vers>
Sep 01 00:48:20 rhel1.example.com systemd[1]: Started Postfix Mail Transport Ag>
lines 1-21/21 (END)
```

Figure 6.21 – Check the status of postfix

In this section, we learned how to set up a null client email server. This is useful for setting up email notifications to be sent from a server to send reports or alerts. We learned how to manipulate `postfix` with the `postconf` command and which variables need to be changed to create a null client.

Setting up email services via Ansible Automation

We are going to set up `postfix` with Ansible Automation using a set of variables and a loop. This will allow us to create the same setup that we got when doing things manually, but done here in an iterative manner using a loop. First, we will set up our playbook directory and then our inventory file. The inventory file can be seen in the following screenshot:

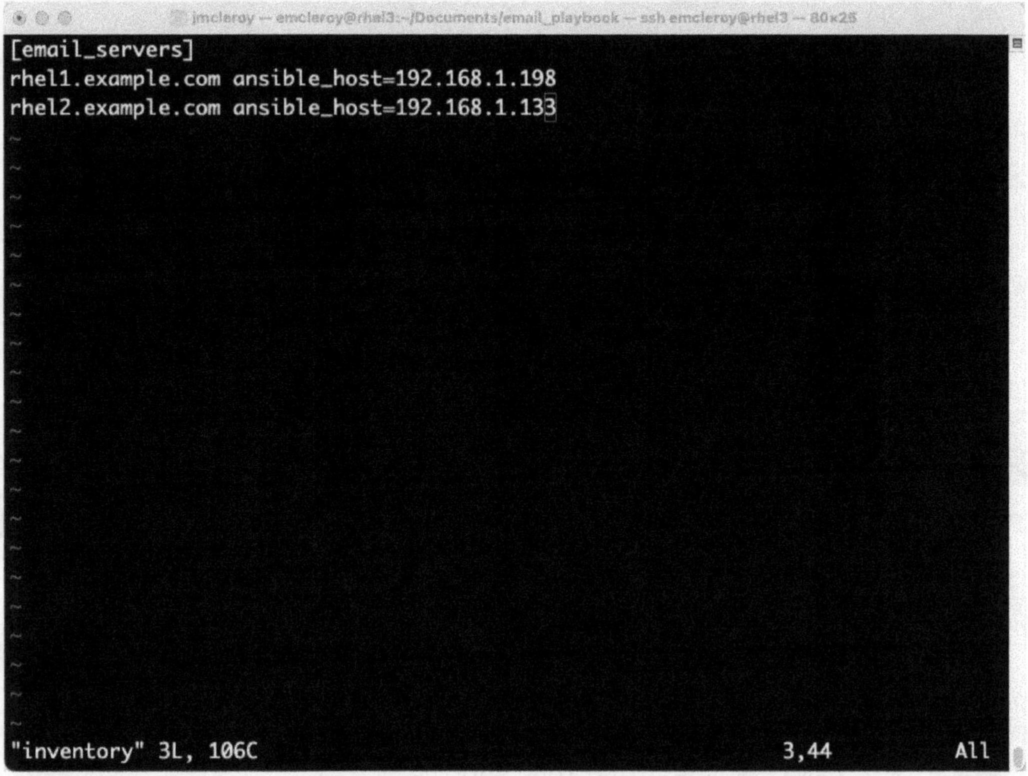

```
jmcleroy — emcleroy@rhel3:~/Documents/email_playbook — ssh emcleroy@rhel3 — 80×25
[email_servers]
rhel1.example.com ansible_host=192.168.1.198
rhel2.example.com ansible_host=192.168.1.133
~
~
~
~
~
~
~
~
~
~
~
~
~
~
~
~
~
~
~
~
"inventory" 3L, 106C                                    3,44              All
```

Figure 6.22 – Inventory for email playbook

We are going to start our playbook with the normal starting format, as seen in the following code:

```
---
- name: Setup null client email server
  hosts: email_servers
  become: true
  become_method: sudo
```

After we have set up the beginning of the playbook, let's create the tasks required to get the playbook to successfully configure a null client. We will use a loop and loop over the variables we provide to the playbook. We can see the playbook in the following code:

```
tasks:
  - name: Install postfix
    package:
      name: postfix
```

```
        state: latest

  - name: Open firewall rules for postfix
    firewalld:
      service: smtp
      state: enabled
      permanent: true

  - name: Reload firewall
    command:
      cmd: firewall-cmd --reload

  - name: Set configuration settings using a loop
    command:
      cmd: postconf -e "{{ item.name }} = {{ item.value }}"
    loop:
      - { name: 'inet_interfaces', value: 'loopback-only' }
      - { name: 'myorigin', value: 'example.com' }
      - { name: 'inet_protocols', value: 'ipv4' }
      - { name: 'mydestination', value: '' }
      - { name: 'mynetworks', value: '127.0.0.0/8' }
      - { name: 'local_transport', value: 'error: no local
delivery' }
      - { name: 'relayhost', value: '[smtp.example.com]' }

  - name: Restart and enable postfix to refresh config
    service:
      name: postfix
      state: restarted
      enabled: true
```

With that done, let's run the playbook with the following command:

```
[emcleroy@rhel3 email_playbook]$ ansible-playbook -i inventory
email_playbook.yml -u emcleroy -k --ask-become -v
```

The truncated output of this playbook when successfully run can be seen in the following screenshot:

```
Bits": "0", "SendSIGHUP": "no", "SendSIGKILL": "yes", "Slice": "system.slice", "
StandardError": "inherit", "StandardInput": "null", "StandardInputData": "", "St
andardOutput": "journal", "StartLimitAction": "none", "StartLimitBurst": "5", "S
tartLimitIntervalUSec": "10s", "StartupBlockIOWeight": "[not set]", "StartupCPUS
hares": "[not set]", "StartupCPUWeight": "[not set]", "StartupIOWeight": "[not s
et]", "StateChangeTimestamp": "Thu 2022-09-01 00:48:20 EDT", "StateChangeTimesta
mpMonotonic": "281389616833", "StateDirectoryMode": "0755", "StatusErrno": "0",
"StopWhenUnneeded": "no", "SubState": "running", "SuccessAction": "none", "Syslo
gFacility": "3", "SyslogLevel": "6", "SyslogLevelPrefix": "yes", "SyslogPriority
": "30", "SystemCallErrorNumber": "0", "TTYReset": "no", "TTYVHangup": "no", "TT
YVTDisallocate": "no", "TasksAccounting": "yes", "TasksCurrent": "3", "TasksMax"
: "11488", "TimeoutStartUSec": "1min 30s", "TimeoutStopUSec": "1min 30s", "Timer
SlackNSec": "50000", "Transient": "no", "Type": "forking", "UID": "[not set]", "
UMask": "0022", "UnitFilePreset": "disabled", "UnitFileState": "enabled", "UtmpM
ode": "init", "WantedBy": "multi-user.target", "WatchdogTimestamp": "Thu 2022-09
-01 00:48:20 EDT", "WatchdogTimestampMonotonic": "281389616832", "WatchdogUSec":
"0"}}


PLAY RECAP ********************************************************************
rhel1.example.com            : ok=6    changed=3    unreachable=0    failed=0    s
kipped=0    rescued=0    ignored=0
rhel2.example.com            : ok=6    changed=3    unreachable=0    failed=0    s
kipped=0    rescued=0    ignored=0

[emcleroy@rhel3 email_playbook]$
```

Figure 6.23 – Successful email setup playbook run

Finally, let's check and confirm our configuration of `mynetworks` to ensure that it is only using `127.0.0.0/8`, as shown in the following screenshot:

jmcleroy — emcleroy@rhel1:~ — ssh emcleroy@rhel1 — 80x25
[emcleroy@rhel1 ~]$ postconf mynetworks
mynetworks = 127.0.0.0/8
[emcleroy@rhel1 ~]$

Figure 6.24 – Confirming the settings changed in the postfix configuration

In this section, we learned how to set up a `postfix` null client using Ansible Automation. This is a great way to set up your mail servers easily using predetermined values that can span multiple hosts. This allows you to quickly set up these servers without having to install and then type in the configuration commands on each and every server.

Summary

In this chapter, we went over how to print from a server and how to set up email hosts to correctly forward emails to a relay using a null server. This allows you to set up subordinate systems to other masters. With this knowledge, you can successfully print from your servers or Linux clients along with sending and receiving mail transmissions. In the next chapter, we are going to work with MariaDB and the MySQL commands that are available for it. A database is a powerful tool that allows you to create backends for web servers and for programs that need a database to run. This will allow you to control your data points with ease. We will do this both manually and via Ansible Automation to make your life easier when setting up multiple databases throughout the course of your career.

7

Databases – Setting Up and Working with MariaDB SQL Databases

In this chapter, we will be discussing databases. In our case, this will involve going into detail about the operation of MariaDB. This is a powerful tool that allows us to store and access data for use by applications, as well as web servers. With the power of a fully functional database behind your project, you will be able to increase the functionality of your application or web server by adding a multitude of items that can be saved and accessed by these products. Things from shopping carts to member information can all be saved using MariaDB. Let's get started with learning more about databases and MariaDB for Red Hat Enterprise Linux 8.1.

In this chapter, we're going to cover the following main topics:

- Getting started with MariaDB for data collection storage
- Installing and configuring MariaDB on RHEL 8 manually
- Installing and configuring MariaDB on RHEL 8 using Ansible Automation

Technical requirements

The technical requirements for this chapter are covered in the following subsection.

Setting up GitHub access

Please refer to the instructions found in *Chapter 1, Block Storage – Learning How to Provision Block Storage on Red Hat Enterprise Linux*, to gain access to GitHub. You will find the Ansible Automation playbooks for this chapter at the following link: `https://github.com/PacktPublishing/Red-Hat-Certified-Specialist-in-Services-Management-and-Automation-EX358-Exam-Guide/tree/main/Chapter07`. Remember these are suggested playbooks and are not the only way you can write them – you can make the playbooks work for you.

You can always change them up using `raw`, `shell`, or `cmd` to achieve the same results but we are demonstrating the best way to accomplish the goals. Also, keep in mind that we are not using the FCQN, which is needed in future versions of Ansible, as that will not be supported in the exam, which covers Ansible 2.9.

Getting started with MariaDB for data collection storage

MariaDB is a database that allows for the relational storage of objects. This allows for key-value pairs to be stored and accessed using MySQL commands. The ability to save and access this information is paramount for platforms such as forums where usernames and passwords need to be correlated together for authentication, for example. We also use this for inventory tracking in many instances, for businesses both big and small. The power of these databases is a necessity that every company relies upon.

Structured Query Language (SQL) is what is normally used by a **relational database management system (RDBMS)** in order to control how a database is configured and accessed. This system provides the means to configure your database in the manner that is needed to function properly, whether it is used by an application or a web server. The more you use relational databases, the more you will understand their significance when it comes to business strategies.

First, we are going to look at manually installing MariaDB and then we'll move on to the automated installation and manipulation of these databases.

Installing and configuring MariaDB on RHEL 8 manually

First, we are going to start the MariaDB installation in order to get started with using this powerful package of tools. We will look up the MariaDB package to get the default version and the profiles available using the command in the following screenshot:

```
[emcleroy@rhel2 ~]$ dnf module list mariadb
Not root, Subscription Management repositories not updated
Red Hat Enterprise Linux 8 for x86_64 - AppStre  33 MB/s |  46 MB    00:01
Red Hat Enterprise Linux 8 for x86_64 - BaseOS   28 MB/s |  51 MB    00:01
Last metadata expiration check: 0:00:08 ago on Sun 04 Sep 2022 09:41:45 AM CDT.
Red Hat Enterprise Linux 8 for x86_64 - AppStream (RPMs)
Name           Stream           Profiles                    Summary
mariadb        10.3 [d][e]      client, server [d], galera  MariaDB Module
mariadb        10.5             client, server [d], galera  MariaDB Module

Hint: [d]efault, [e]nabled, [x]disabled, [i]nstalled
[emcleroy@rhel2 ~]$
```

Figure 7.1 – Module list of the server package instead of the client package

As shown in the preceding screenshot, we can see that MariaDB 10.3 is the default, and we can see the profile of the server, which is what we want to install. We will install the MariaDB server as seen in the following screenshot:

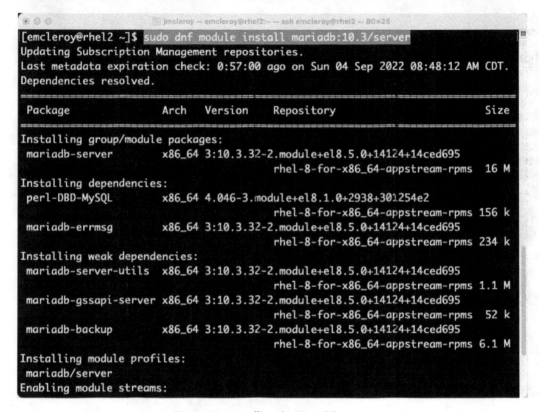

Figure 7.2 – Installing the MariaDB server

We will then start and enable the MariaDB service as shown in the following screenshot:

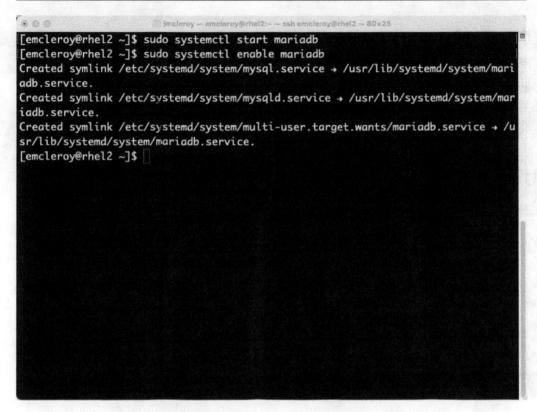

Figure 7.3 – Starting and enabling the MariaDB service

We will then open the ports for MySQL, which, as discussed earlier, is how we interact with MariaDB. Opening the firewall rules can be seen in the following screenshot:

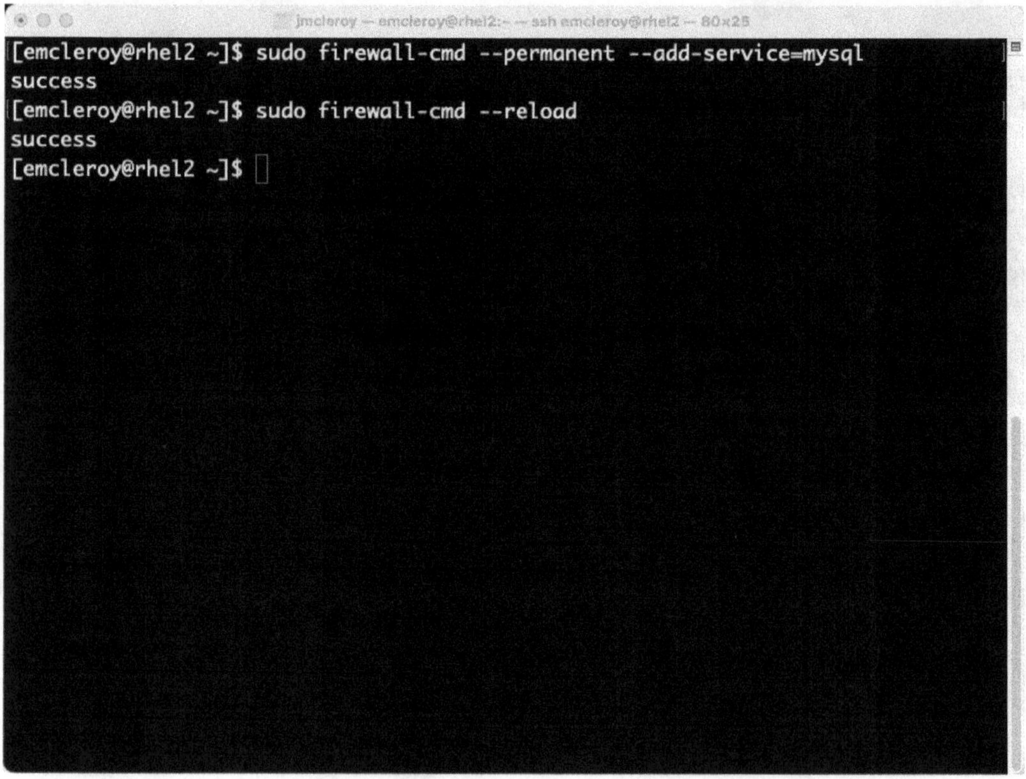
jmcleroy — emcleroy@rhel2:~ — ssh emcleroy@rhel2 — 80×25
[emcleroy@rhel2 ~]$ sudo firewall-cmd --permanent --add-service=mysql
success
[emcleroy@rhel2 ~]$ sudo firewall-cmd --reload
success
[emcleroy@rhel2 ~]$

Figure 7.4 – Opening the firewall rules for MariaDB

Once we have the package enabled and started and the firewall open, we can begin to use our database server for data collection or application backing.

Next, let's look at the options in the configuration file for controlling network connectivity. We have learned in the previous chapters that controlling network connectivity is key for many purposes. We will look at `/etc/my.cnf.d/mariadb-server.cnf` and pay close attention to a few items to go over, but first, a sample of this file can be seen in the following screenshot:

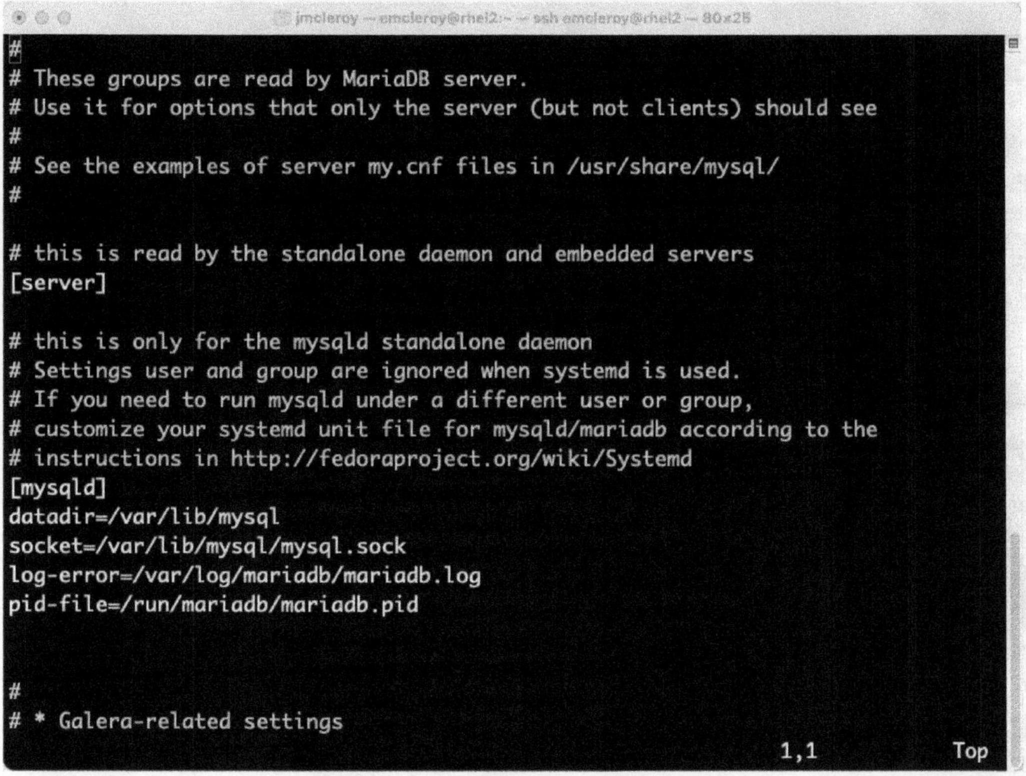

```
jmcleroy — pmcleroy@rhel2:~ — ssh pmcleroy@rhel2 — 80×25
#
# These groups are read by MariaDB server.
# Use it for options that only the server (but not clients) should see
#
# See the examples of server my.cnf files in /usr/share/mysql/
#

# this is read by the standalone daemon and embedded servers
[server]

# this is only for the mysqld standalone daemon
# Settings user and group are ignored when systemd is used.
# If you need to run mysqld under a different user or group,
# customize your systemd unit file for mysqld/mariadb according to the
# instructions in http://fedoraproject.org/wiki/Systemd
[mysqld]
datadir=/var/lib/mysql
socket=/var/lib/mysql/mysql.sock
log-error=/var/log/mariadb/mariadb.log
pid-file=/run/mariadb/mariadb.pid


#
# * Galera-related settings
                                                        1,1         Top
```

Figure 7.5 – The truncated mariadb-server.cnf file

If you look at the Galera-related settings, you will see an example of some of the items we want to pay attention to, as shown in the following screenshot:

```
# * Galera-related settings
#
[galera]
# Mandatory settings
#wsrep_on=ON
#wsrep_provider=
#wsrep_cluster_address=
#binlog_format=row
#default_storage_engine=InnoDB
#innodb_autoinc_lock_mode=2
#
# Allow server to accept connections on all interfaces.
#
#bind-address=0.0.0.0
#
# Optional setting
#wsrep_slave_threads=1
#innodb_flush_log_at_trx_commit=0

# this is only for embedded server
[embedded]

# This group is only read by MariaDB servers, not by MySQL.
# If you use the same .cnf file for MySQL and MariaDB,
                                                  42,0-1         74%
```

Figure 7.6 – The mariadb-server.cnf file (continued)

Here, we can see `bind-address`, which we want to move up to the `[mysqld]` section. Next, we want to add `skip-networking=1` so that connections are only allowed on the server. If `skip-networking` is set to `0`, it will listen for connections outside of `localhost`.

For our use, we are going to limit the bind address to IPv4 and we are going to set `skip-networking=1` by adding a line of code to lock out all connections started from outside of the server. These settings configured within the file can be seen in the following screenshot:

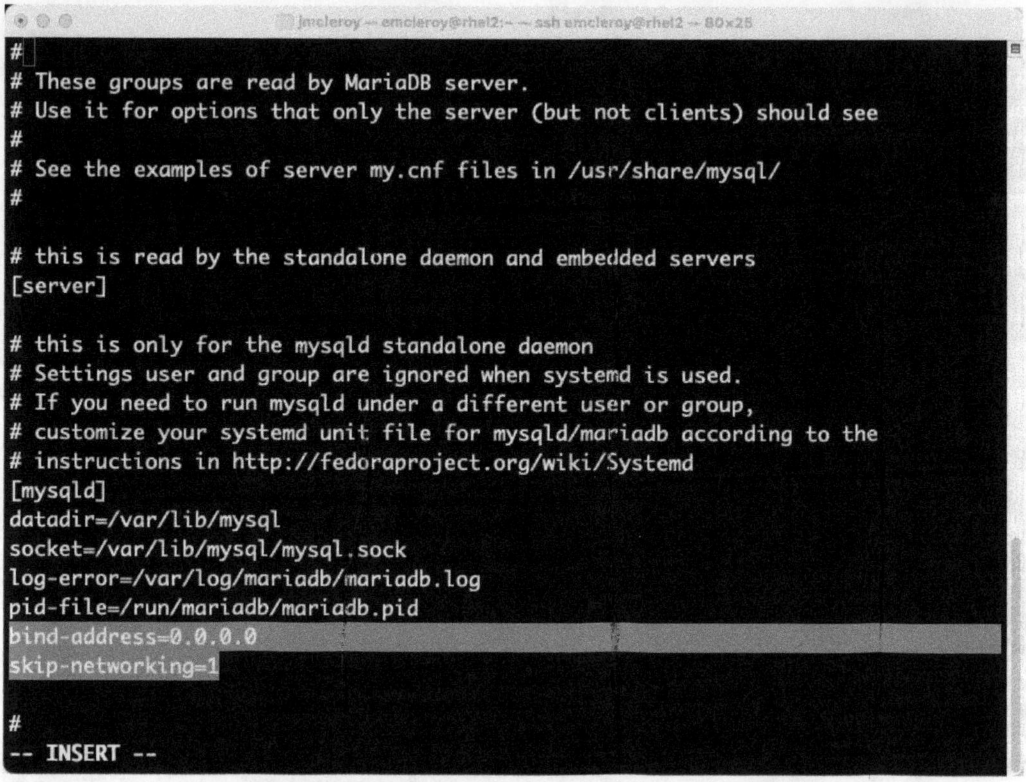

```
jmcleroy — emcleroy@rhel2:~ — ssh emcleroy@rhel2 — 80×25
#
# These groups are read by MariaDB server.
# Use it for options that only the server (but not clients) should see
#
# See the examples of server my.cnf files in /usr/share/mysql/
#

# this is read by the standalone daemon and embedded servers
[server]

# this is only for the mysqld standalone daemon
# Settings user and group are ignored when systemd is used.
# If you need to run mysqld under a different user or group,
# customize your systemd unit file for mysqld/mariadb according to the
# instructions in http://fedoraproject.org/wiki/Systemd
[mysqld]
datadir=/var/lib/mysql
socket=/var/lib/mysql/mysql.sock
log-error=/var/log/mariadb/mariadb.log
pid-file=/run/mariadb/mariadb.pid
bind-address=0.0.0.0
skip-networking=1

#
-- INSERT --
```

Figure 7.7 – The updated mariadb-server.cnf file with skip-networking

Once the settings have been added, ensure that you restart the MariaDB service using the following command:

```
[emcleroy@rhel1 ~]$ sudo systemctl restart mysql
```

Next, we are going to work on manipulating data in the database. We will use MySQL tools to perform these actions. We will start by logging in to the database using the command shown in the following screenshot:

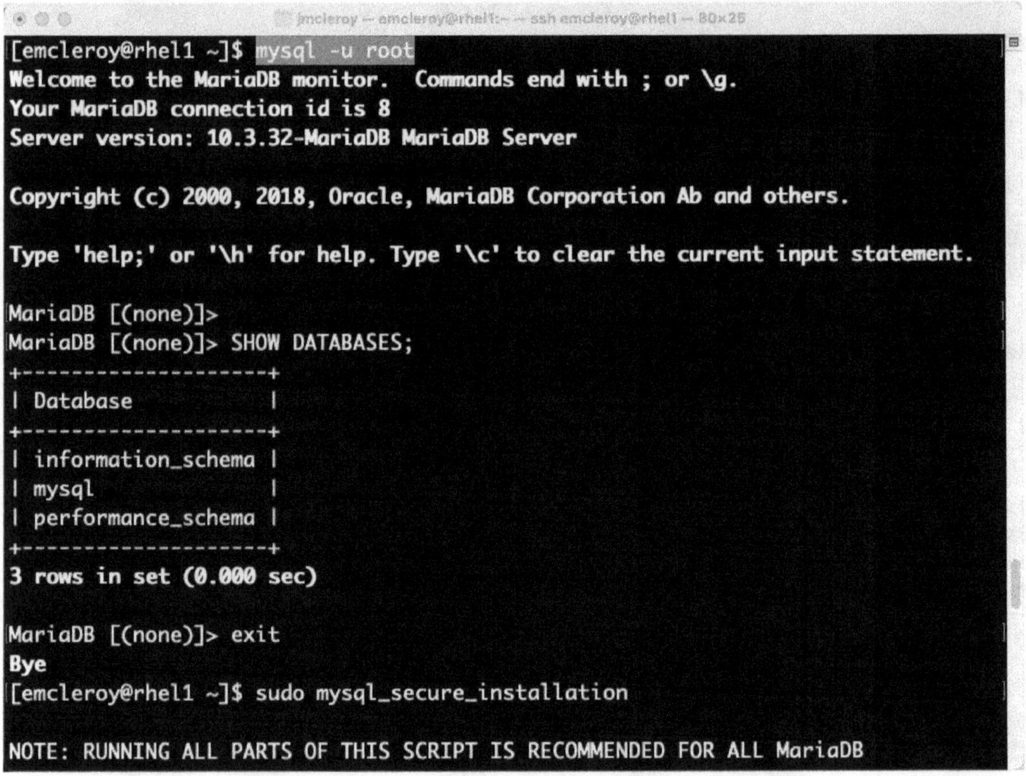

```
[emcleroy@rhel1 ~]$ mysql -u root
Welcome to the MariaDB monitor.  Commands end with ; or \g.
Your MariaDB connection id is 8
Server version: 10.3.32-MariaDB MariaDB Server

Copyright (c) 2000, 2018, Oracle, MariaDB Corporation Ab and others.

Type 'help;' or '\h' for help. Type '\c' to clear the current input statement.

MariaDB [(none)]>
MariaDB [(none)]> SHOW DATABASES;
+--------------------+
| Database           |
+--------------------+
| information_schema |
| mysql              |
| performance_schema |
+--------------------+
3 rows in set (0.000 sec)

MariaDB [(none)]> exit
Bye
[emcleroy@rhel1 ~]$ sudo mysql_secure_installation

NOTE: RUNNING ALL PARTS OF THIS SCRIPT IS RECOMMENDED FOR ALL MariaDB
```

Figure 7.8 – Logging in to the mysql database for the first time

We are now going to secure the database by running the following command to secure the MySQL database installation:

```
[emcleroy@rhel2 ~]$ sudo mysql_secure_installation
```

This command removes things such as the test database and the anonymous user and sets a password for root accounts, among others. We can see the setup in action in the following screenshot:

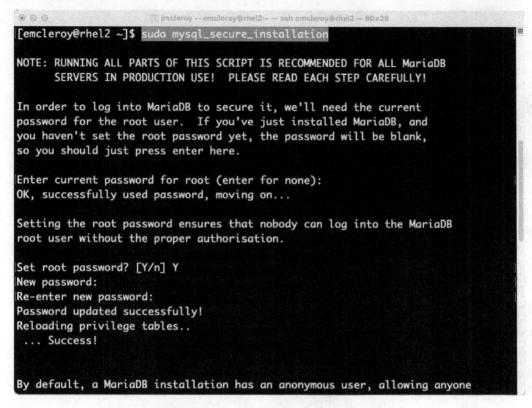

```
[emcleroy@rhel2 ~]$ sudo mysql_secure_installation

NOTE: RUNNING ALL PARTS OF THIS SCRIPT IS RECOMMENDED FOR ALL MariaDB
      SERVERS IN PRODUCTION USE!  PLEASE READ EACH STEP CAREFULLY!

In order to log into MariaDB to secure it, we'll need the current
password for the root user.  If you've just installed MariaDB, and
you haven't set the root password yet, the password will be blank,
so you should just press enter here.

Enter current password for root (enter for none):
OK, successfully used password, moving on...

Setting the root password ensures that nobody can log into the MariaDB
root user without the proper authorisation.

Set root password? [Y/n] Y
New password:
Re-enter new password:
Password updated successfully!
Reloading privilege tables..
 ... Success!


By default, a MariaDB installation has an anonymous user, allowing anyone
```

Figure 7.9 – Securing the MariaDB installation

Next, we are going to look at the databases and you will notice there are some default databases. These are used to handle how MySQL works within MariaDB, as shown in the following screenshot:

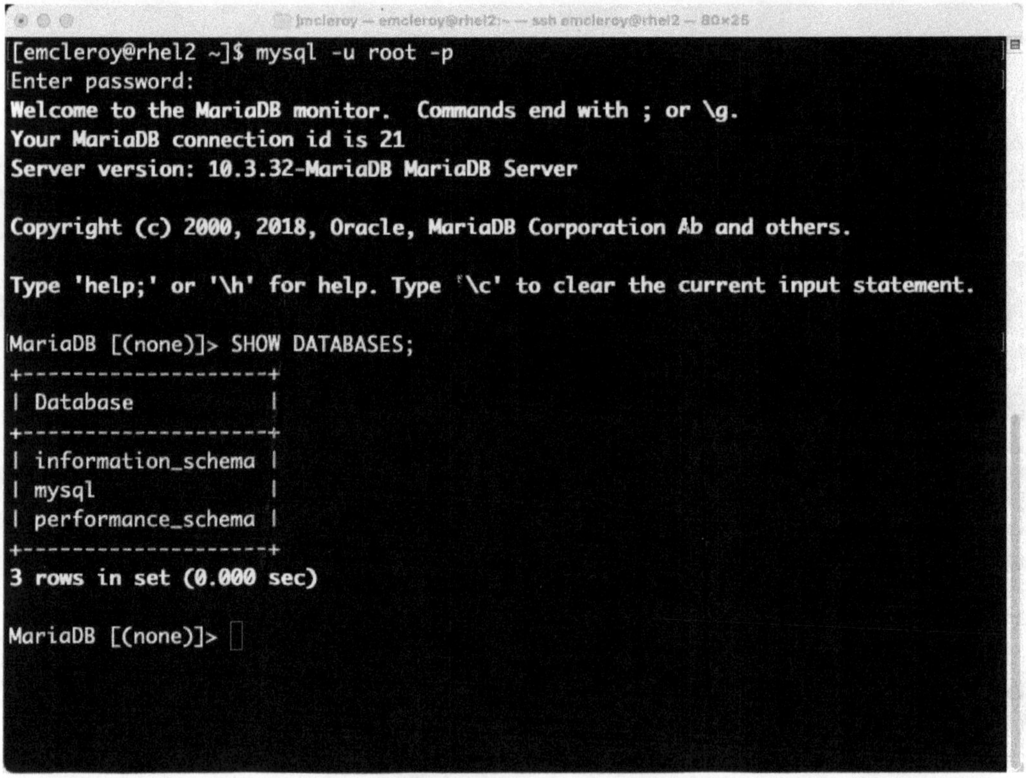

Figure 7.10 – Logging in using the newly set password and showing the databases

In order to use a database, we will use the USE command, such as USE information_schema;, in order to access and manipulate that database. Next, we will use the database creation command and then use that database to manipulate data. We will create this new personnel database and make it ready for use as shown in the following screenshot:

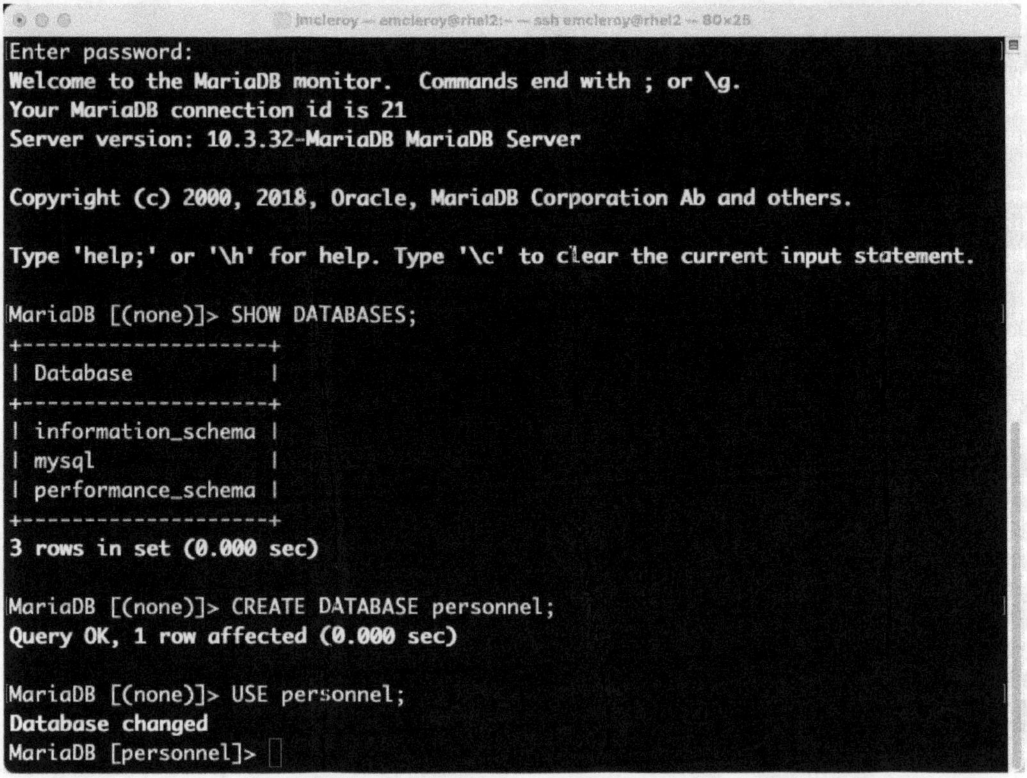

```
Enter password:
Welcome to the MariaDB monitor.  Commands end with ; or \g.
Your MariaDB connection id is 21
Server version: 10.3.32-MariaDB MariaDB Server

Copyright (c) 2000, 2018, Oracle, MariaDB Corporation Ab and others.

Type 'help;' or '\h' for help. Type '\c' to clear the current input statement.

MariaDB [(none)]> SHOW DATABASES;
+--------------------+
| Database           |
+--------------------+
| information_schema |
| mysql              |
| performance_schema |
+--------------------+
3 rows in set (0.000 sec)

MariaDB [(none)]> CREATE DATABASE personnel;
Query OK, 1 row affected (0.000 sec)

MariaDB [(none)]> USE personnel;
Database changed
MariaDB [personnel]>
```

Figure 7.11 – Using the personnel database

After that, we will want to inspect the database and see whether anything is in it. For this, we will switch to the personnel database. We can identify that the database is currently empty in the following screenshot:

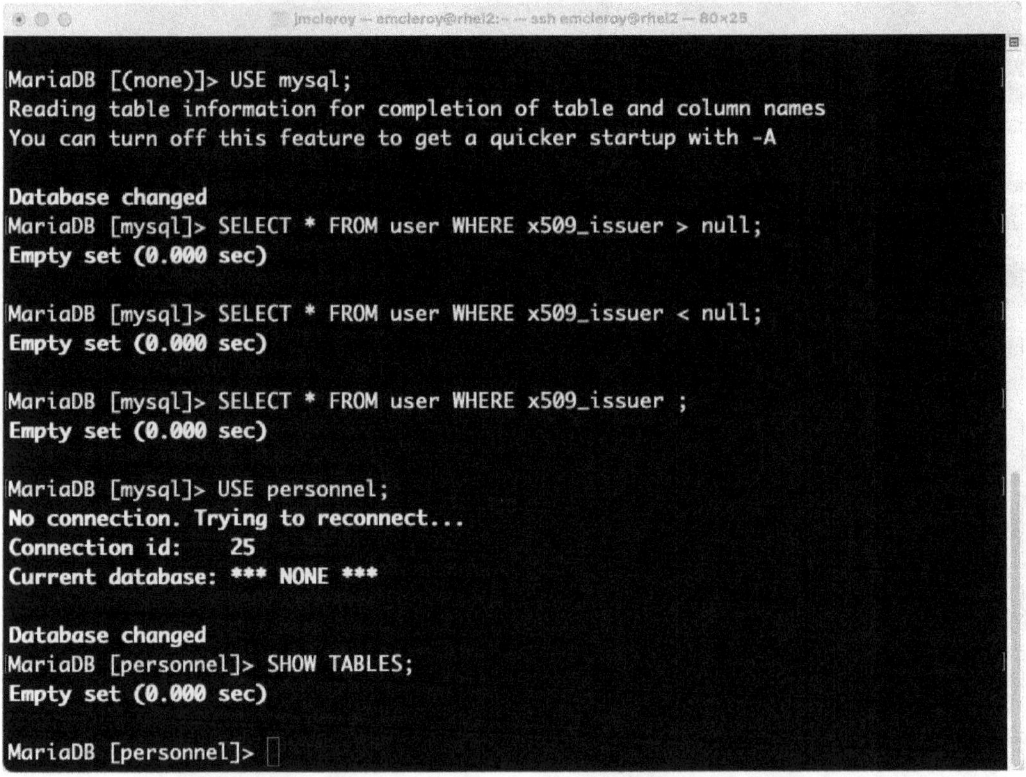

```
MariaDB [(none)]> USE mysql;
Reading table information for completion of table and column names
You can turn off this feature to get a quicker startup with -A

Database changed
MariaDB [mysql]> SELECT * FROM user WHERE x509_issuer > null;
Empty set (0.000 sec)

MariaDB [mysql]> SELECT * FROM user WHERE x509_issuer < null;
Empty set (0.000 sec)

MariaDB [mysql]> SELECT * FROM user WHERE x509_issuer ;
Empty set (0.000 sec)

MariaDB [mysql]> USE personnel;
No connection. Trying to reconnect...
Connection id:    25
Current database: *** NONE ***

Database changed
MariaDB [personnel]> SHOW TABLES;
Empty set (0.000 sec)

MariaDB [personnel]> 
```

Figure 7.12 – Showing the tables are empty within the personnel database

After we have created the database, we are going to create a table for the database using the command shown in the following screenshot:

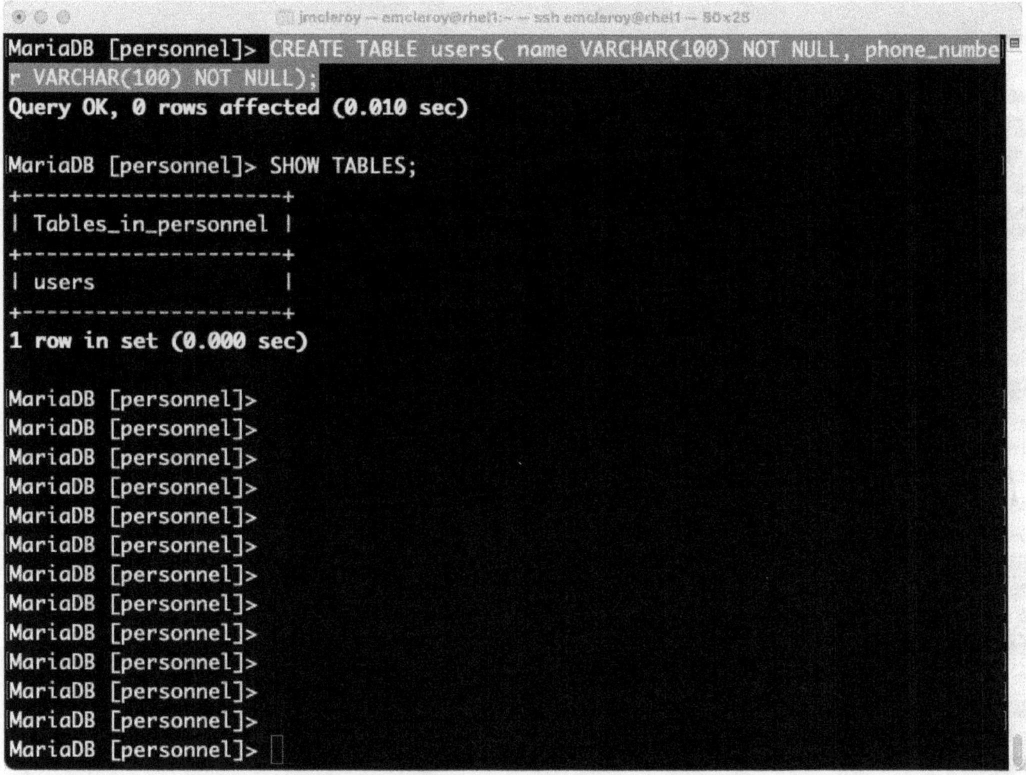

Figure 7.13 – Creating tables in the personnel database

Next, we will be adding information to the name and phone number columns so that we have something to search through. We can see the commands to add to the database table in the following screenshot:

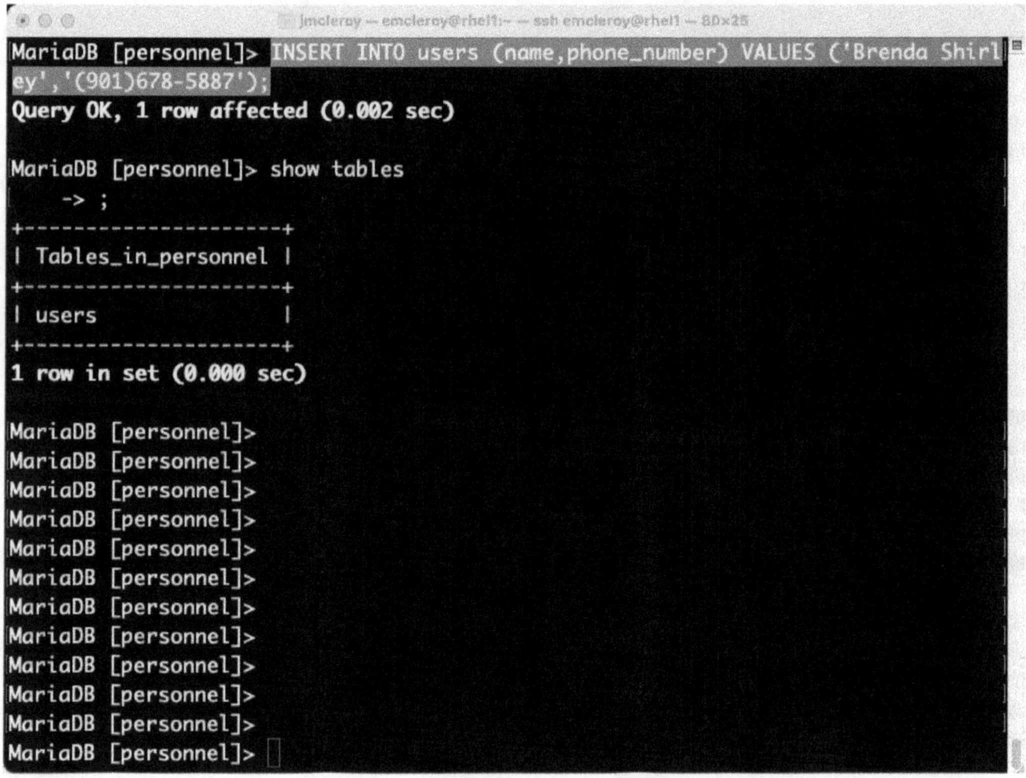

Figure 7.14 – Creating data within the table of the personnel database

You can also search for items within tables using the command shown in the following screenshot:

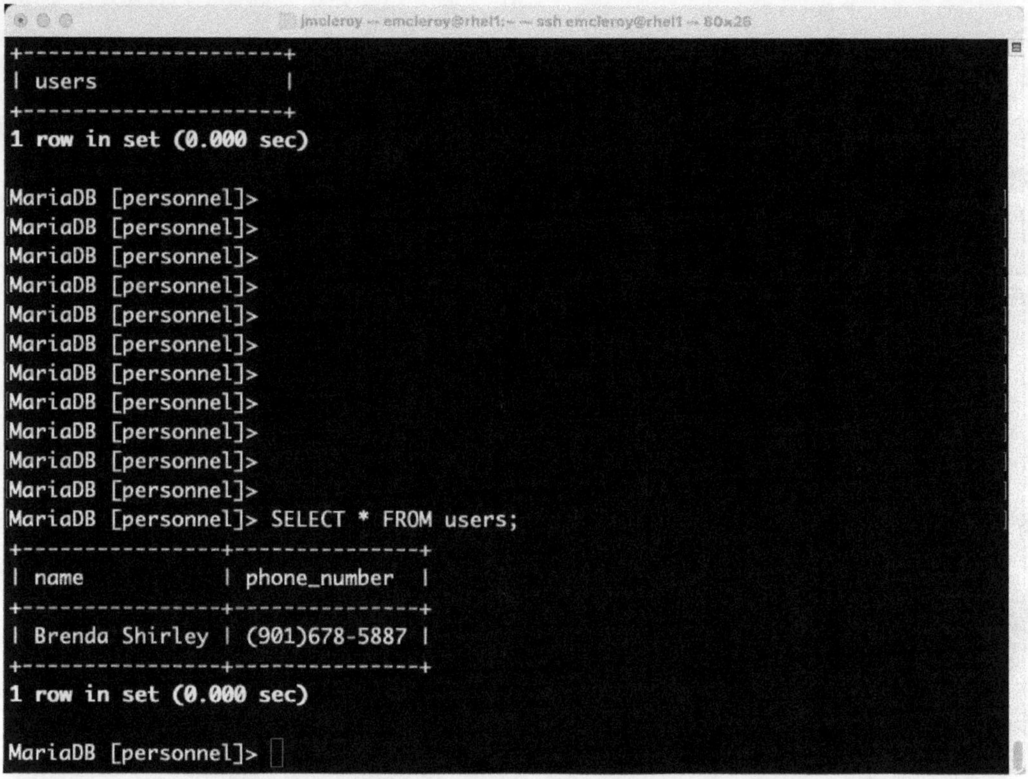

Figure 7.15 – Searching the database using known information

The SELECT command can be used with the WHERE variable, as in the following command, to look for a specific column within a database:

```
SELECT * FROM users WHERE phone_number = '(901)678-5887';
```

Next, we are going to talk about taking a backup of your database. This will allow you to restore information that you have backed up recently using the mysqldump command. More about the mysqldump command can be found in the man page for it, as shown in the following screenshot:

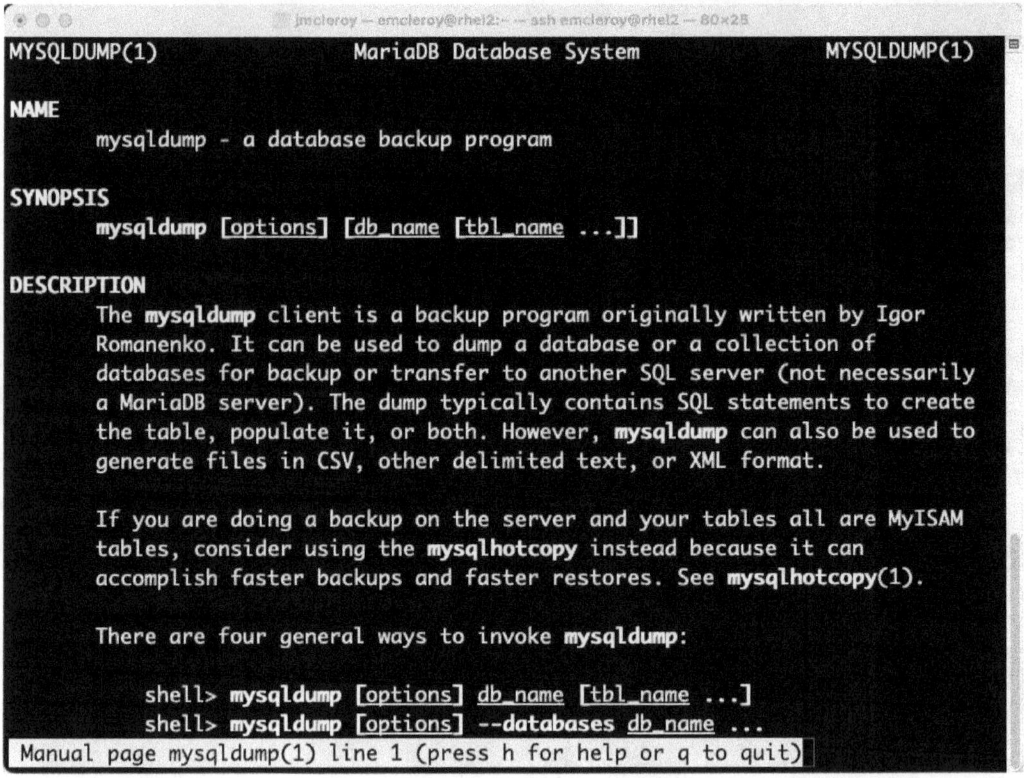

Figure 7.16 – mysqldump man page

We will use the following command to back up the database:

```
[emcleroy@rhel2 ~]$ mysqldump -u root -p personnel >/tmp/
mariadb.dump
```

This command execution can be seen in the following screenshot:

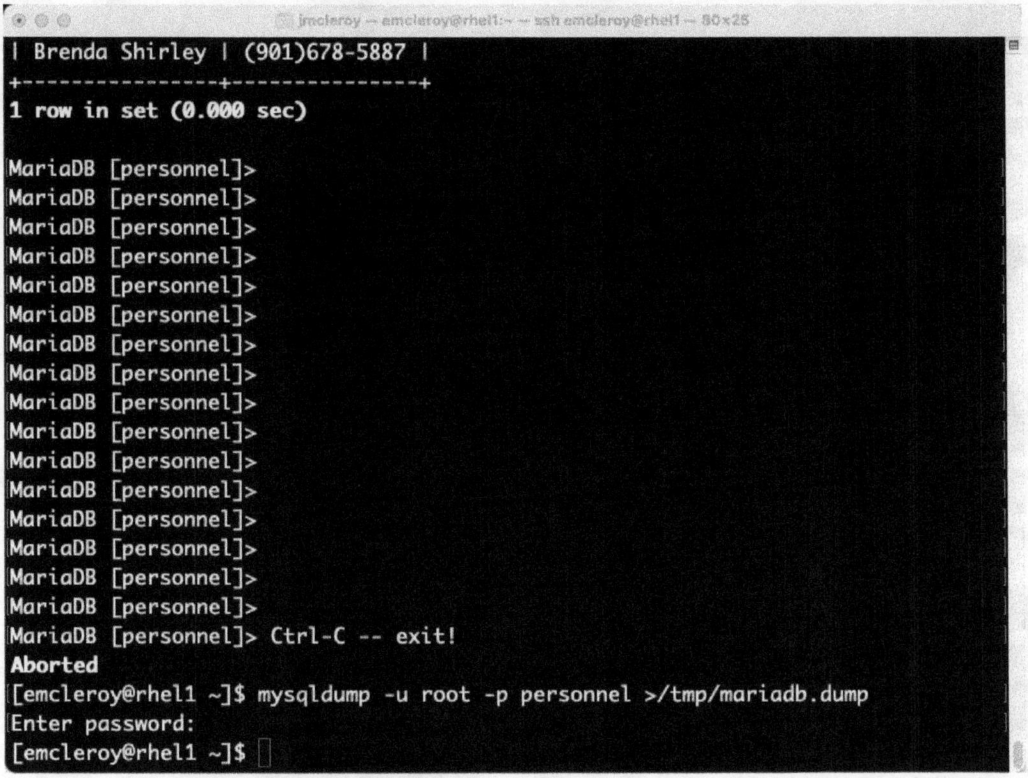

```
| Brenda Shirley | (901)678-5887 |
+----------------+----------------+
1 row in set (0.000 sec)

MariaDB [personnel]>
MariaDB [personnel]>
MariaDB [personnel]>
MariaDB [personnel]>
MariaDB [personnel]>
MariaDB [personnel]>
MariaDB [personnel]>
MariaDB [personnel]>
MariaDB [personnel]>
MariaDB [personnel]>
MariaDB [personnel]>
MariaDB [personnel]>
MariaDB [personnel]>
MariaDB [personnel]>
MariaDB [personnel]>
MariaDB [personnel]>
MariaDB [personnel]> Ctrl-C -- exit!
Aborted
[emcleroy@rhel1 ~]$ mysqldump -u root -p personnel >/tmp/mariadb.dump
Enter password:
[emcleroy@rhel1 ~]$
```

Figure 7.17 – Creating a backup of the personnel database

Next, we are going to restore from the backup. We will use the following command to choose the backup file we want to restore and restore it to our database:

```
[emcleroy@rhel2 ~]$ mysql -u root -p personnel < /tmp/mariadb.
dump
```

Let's test this by dropping the users table as shown in the following screenshot and then restoring the database:

```
[emcleroy@rhel1 ~]$ mysql -u root -p
Enter password:
Welcome to the MariaDB monitor.  Commands end with ; or \g.
Your MariaDB connection id is 20
Server version: 10.3.32-MariaDB MariaDB Server

Copyright (c) 2000, 2018, Oracle, MariaDB Corporation Ab and others.

Type 'help;' or '\h' for help. Type '\c' to clear the current input statement.

MariaDB [(none)]> use personnel;
Reading table information for completion of table and column names
You can turn off this feature to get a quicker startup with -A

Database changed
MariaDB [personnel]> drop table users;
Query OK, 0 rows affected (0.006 sec)

MariaDB [personnel]> show tables
    -> ;
Empty set (0.000 sec)

MariaDB [personnel]>
```

Figure 7.18 – Removing tables and checking that they are missing before restoring them

Now, we are going to restore the database and show that the table has been recreated, as shown in the following screenshot:

```
[emcleroy@rhel1 ~]$ mysql -u root -p personnel < /tmp/mariadb.dump
Enter password:
[emcleroy@rhel1 ~]$ mysql -u root -p
Enter password:
Welcome to the MariaDB monitor.  Commands end with ; or \g.
Your MariaDB connection id is 22
Server version: 10.3.32-MariaDB MariaDB Server

Copyright (c) 2000, 2018, Oracle, MariaDB Corporation Ab and others.

Type 'help;' or '\h' for help. Type '\c' to clear the current input statement.

MariaDB [(none)]> use personnel;
Reading table information for completion of table and column names
You can turn off this feature to get a quicker startup with -A

Database changed
MariaDB [personnel]> show tables;
+----------------------+
| Tables_in_personnel  |
+----------------------+
| users                |
+----------------------+
1 row in set (0.000 sec)
```

Figure 7.19 – Restoring the database and showing that the tables are now back

Once you load this information back into your database, you can use show commands to gather information for anything needed.

Finally, we are going to show how to control user access to your systems. We are going to add users with different rights to the databases. Adding these users can be done with the following command in this screenshot:

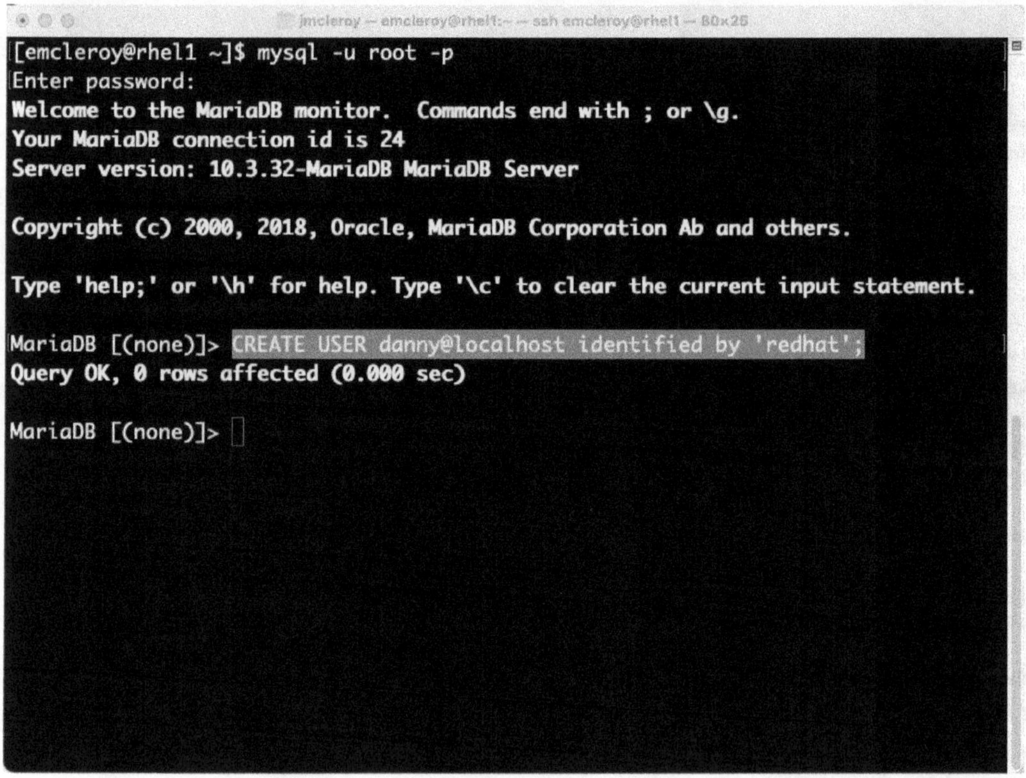

```
[emcleroy@rhel1 ~]$ mysql -u root -p
Enter password:
Welcome to the MariaDB monitor.  Commands end with ; or \g.
Your MariaDB connection id is 24
Server version: 10.3.32-MariaDB MariaDB Server

Copyright (c) 2000, 2018, Oracle, MariaDB Corporation Ab and others.

Type 'help;' or '\h' for help. Type '\c' to clear the current input statement.

MariaDB [(none)]> CREATE USER danny@localhost identified by 'redhat';
Query OK, 0 rows affected (0.000 sec)

MariaDB [(none)]>
```

Figure 7.20 – Creating users for the databases

In the preceding screenshot, you can see we created danny@localhost, specifying that this is a local user and that it is identified by, referring to the password for the user, redhat, in our case. Next, we are going to add privileges to danny so that he can make changes to the personnel database. The commands for adding privileges can be seen in the following commands within the screenshot:

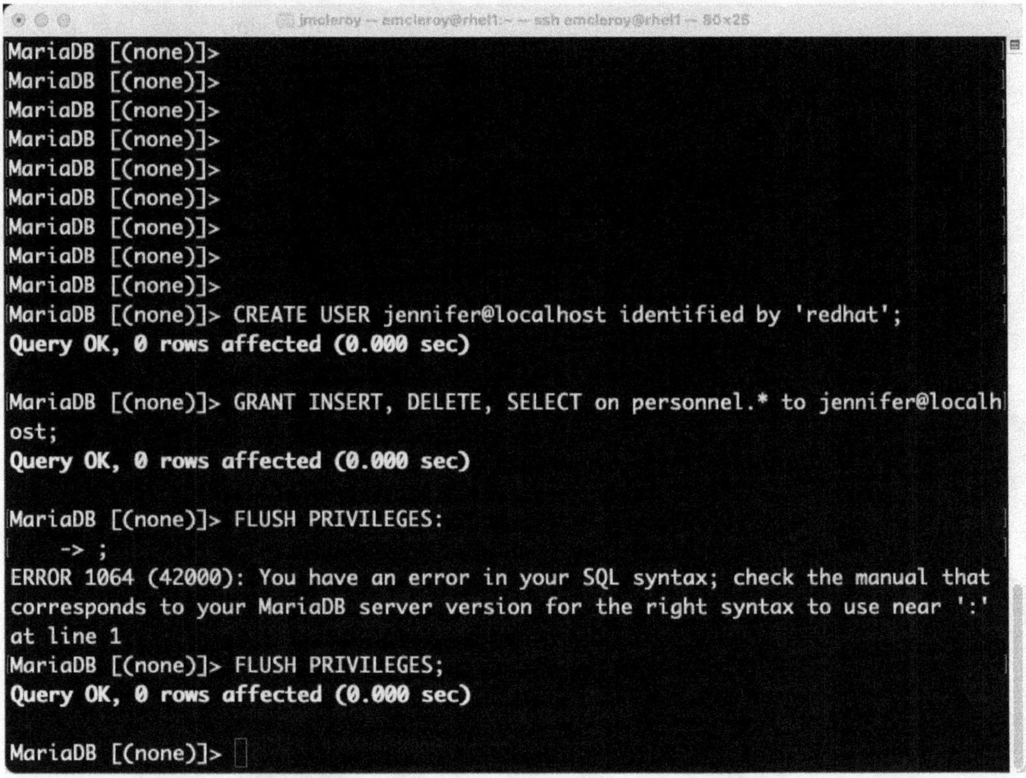

```
MariaDB [(none)]>
MariaDB [(none)]>
MariaDB [(none)]>
MariaDB [(none)]>
MariaDB [(none)]>
MariaDB [(none)]>
MariaDB [(none)]>
MariaDB [(none)]>
MariaDB [(none)]>
MariaDB [(none)]> CREATE USER jennifer@localhost identified by 'redhat';
Query OK, 0 rows affected (0.000 sec)

MariaDB [(none)]> GRANT INSERT, DELETE, SELECT on personnel.* to jennifer@localh
ost;
Query OK, 0 rows affected (0.000 sec)

MariaDB [(none)]> FLUSH PRIVILEGES:
    -> ;
ERROR 1064 (42000): You have an error in your SQL syntax; check the manual that
corresponds to your MariaDB server version for the right syntax to use near ':'
at line 1
MariaDB [(none)]> FLUSH PRIVILEGES;
Query OK, 0 rows affected (0.000 sec)

MariaDB [(none)]>
```

Figure 7.21 – Setting privileges for a new user created

We have also created jennifer as a user and provided her with privileges to alter the database. This allows jennifer to insert, delete, and select items within the personnel database. Now, we can use jennifer to add items to the database.

Installing and configuring MariaDB on RHEL 8 using Ansible Automation

We will start with the inventory file to set up the MariaDB servers and configure them with users. We will then go into manipulating data within the created MariaDB system using loops to add things such as tables and database entries. We will then use Ansible Automation to show how to retrieve information from the MariaDB database so that you can provide information from your database to end users. Please ensure that the server you are running this playbook against has not had MariaDB installed previously, or you will receive errors when trying to set the root password.

First, we are going to start with an inventory file, which can be seen in the following screenshot:

Figure 7.22 – Ansible Automation inventory file

From there, we are going to create the `ansible.cfg` file within our directory so that we can skip `host_key_checking`, as it causes issues with connectivity. We can see the Ansible file in the following screenshot:

```
[defaults]
host_key_checking: false

-- INSERT --                                          2,25              All
```

Figure 7.23 – Ansible Automation ansible.cfg file

After we have those files, we will create the playbook file of `mariadb_server.yml` and start it as normal, as shown in the following code:

```
---
- name: MariaDB install and configuration
  hosts: mariadb_servers
  become: true
  become_method: sudo
```

Next, we are going to install MariaDB and set up the firewall rules as well. Installing the package and starting and enabling the service, as well as setting the firewall rules, can be seen in the following code:

```
tasks:
  - name: Install MariaDB server
    package:
      name: '@mariadb:10.3/server'
      state: present

  - name: Install MariaDB client
```

```
      package:
        name: mariadb
        state: latest

    - name: Start and enable MariaDB
      service:
        name: mariadb
        state: started
        enabled: true

    - name: Open firewall rules for MariaDB
      firewalld:
        service: mysql
        permanent: true
        state: enabled

    - name: Reload firewall
      command:
        cmd: firewall-cmd --reload
```

After we have the code for installing the server, we will move on to securing the MariaDB installation to enable it and open up the firewall rules. We will start by configuring the root password. We can see that we have configured that in the following code and marked it with `no_log: true` in order to keep the system from outputting this in the log files:

```
    - name: Install PyMySQL
      package:
        name: python3-PyMySQL
        state: present
    - name: Set root password for MariaDB
      mysql_user:
        name: root
        host_all: true
        update_password: always
        password: redhat
      no_log: true
      ignore_errors: true
```

Next, we are going to copy a configuration file to `root` so that we can log in and make changes to the database without having to show the user and password. We can also vault this file, but for our purposes, we are simply going to copy it. The file can be seen in the following screenshot:

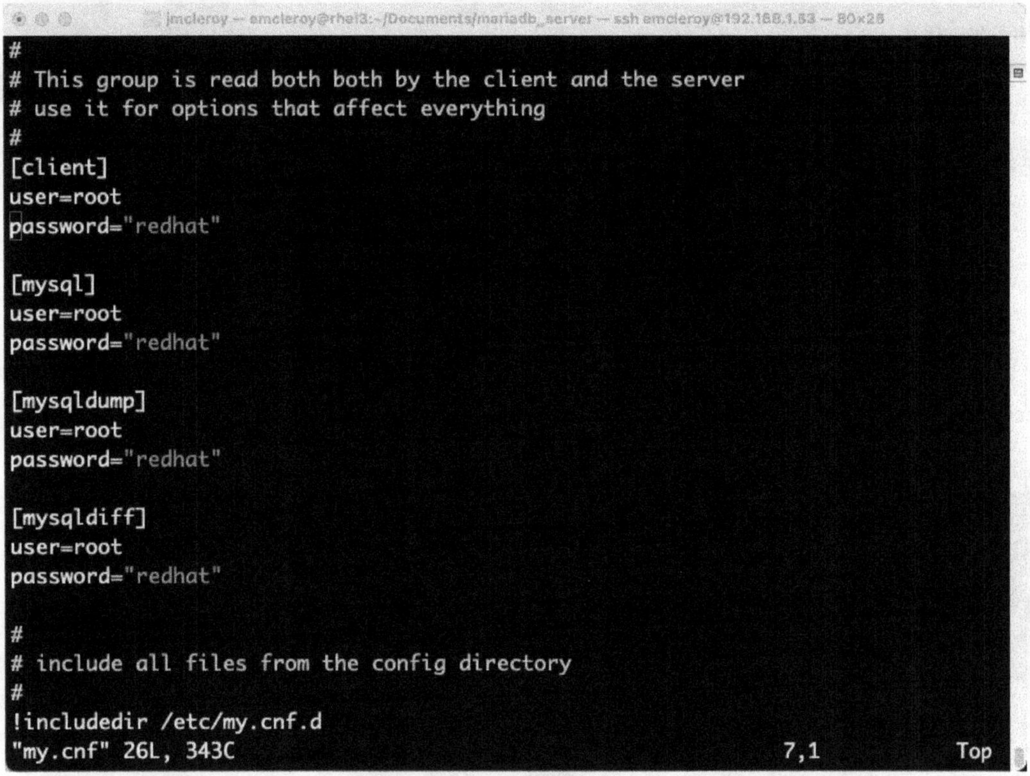

```
#
# This group is read both both by the client and the server
# use it for options that affect everything
#
[client]
user=root
password="redhat"

[mysql]
user=root
password="redhat"

[mysqldump]
user=root
password="redhat"

[mysqldiff]
user=root
password="redhat"

#
# include all files from the config directory
#
!includedir /etc/my.cnf.d
"my.cnf" 26L, 343C                                    7,1              Top
```

Figure 7.24 – The my.cnf file to provide access for configuring databases

The following code will be used to move the file to the needed location on the server for use later on with the modules to add and change things on the system:

```
- name: Copy user and password to home
  copy:
    src: "{{ playbook_dir }}/my.cnf"
    dest: /root/.my.cnf
```

After that, we will remove the anonymous user so that only authenticated users can access the databases. We can see this removal of the anonymous user in the following code:

```
- name: Delete anonymous user
  mysql_user:
```

```
        name: ''
        host_all: yes
        state: absent
```

Next, we are going to create a database called `employees`. We will update this database using a loop so that it is configured with multiple employees within the database. We can see how to create the database and tables and fill in the data in the following code:

```
- name: Add employees database
  mysql_db:
      name: employees
      state: present
      config_file: /root/.my.cnf
```

Next, we will create the `insert.sql` file in order to import the needed database items. This can be seen in the following screenshot:

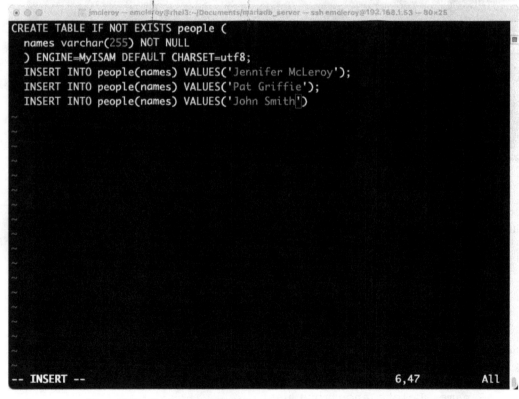

Figure 7.25 – The insert.sql file to import data into the employee database

After creating the database, we will now create a table within the database with the name of people, as shown in the following code, using the `insert.sql` file to provide the information for the table and data:

```
- name: Copy insert.sql file to tmp
  copy:
    src: "{{ playbook_dir }}/insert.sql"
    dest: /tmp/insert.sql

- name: Import the insert.sql information into the database
  mysql_db:
    name: employees
    state: import
    target: /tmp/insert.sql
    config_file: /root/.my.cnf
```

Next, we are going to create a user for the databases and provide them with permissions to the newly created database. We can see how we create users with set privileges in the following code:

```
- name: Add user
  mysql_user:
    name: david
    password: redhat
    update_password: on_create
    state: present
  no_log: true

- name: Provide user with privileges
  mysql_user:
    name: david
    host: localhost
    priv: 'people.*:INSERT,UPDATE,SELECT'
    state: present
```

Now that we have all of this set, we are going to run the playbook. We will use the following command to run the playbook again, assuming you do not have keys set up:

```
[emcleroy@rhel3 mariadb_server]$ ansible-playbook -i inventory
mariadb_server.yml -u emcleroy -k --ask-become -v
```

The following screenshot shows a successful Ansible Automation playbook run:

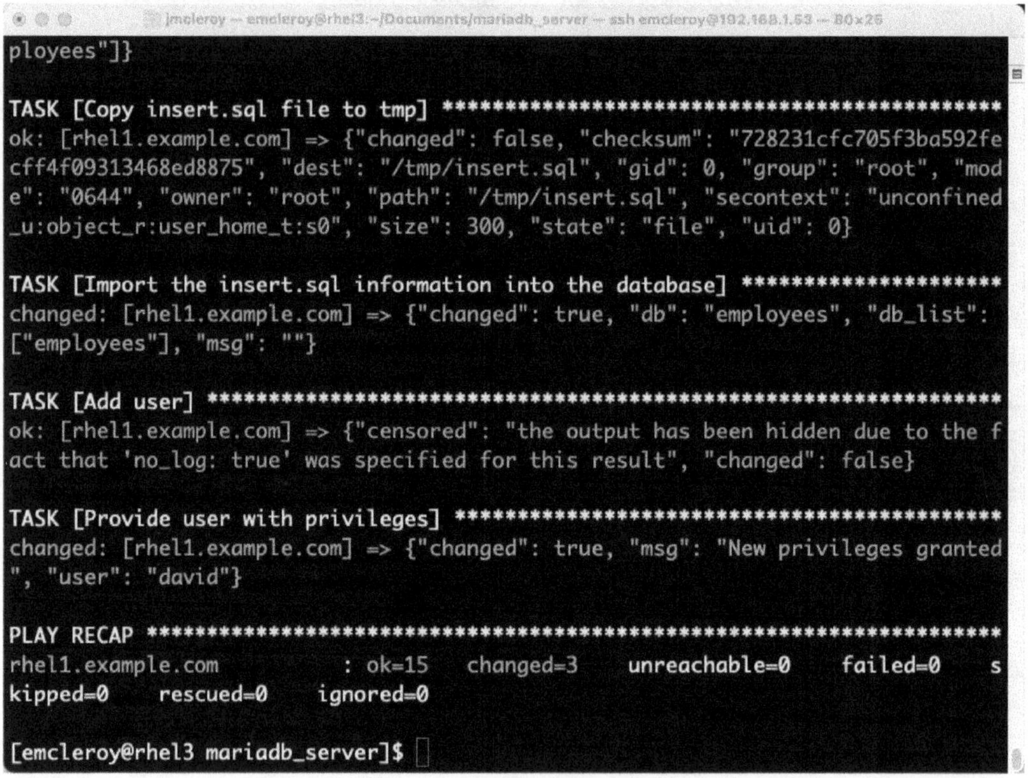
jmcleroy — emcleroy@rhel3:~/Documents/mariadb_server — ssh emcleroy@192.168.1.53 — 80×26
ployees"]}
TASK [Copy insert.sql file to tmp] **
ok: [rhel1.example.com] => {"changed": false, "checksum": "728231cfc705f3ba592fe
cff4f09313468ed8875", "dest": "/tmp/insert.sql", "gid": 0, "group": "root", "mod
e": "0644", "owner": "root", "path": "/tmp/insert.sql", "secontext": "unconfined
_u:object_r:user_home_t:s0", "size": 300, "state": "file", "uid": 0}
TASK [Import the insert.sql information into the database] ********************
changed: [rhel1.example.com] => {"changed": true, "db": "employees", "db_list":
["employees"], "msg": ""}
TASK [Add user] ***
ok: [rhel1.example.com] => {"censored": "the output has been hidden due to the f
act that 'no_log: true' was specified for this result", "changed": false}
TASK [Provide user with privileges] ***
changed: [rhel1.example.com] => {"changed": true, "msg": "New privileges granted
", "user": "david"}
PLAY RECAP **
rhel1.example.com : ok=15 changed=3 unreachable=0 failed=0 s
kipped=0 rescued=0 ignored=0
[emcleroy@rhel3 mariadb_server]$

Figure 7.26 – Successful playbook run

After a successful playbook run, you can see the information has been added, as shown in the following screenshot:

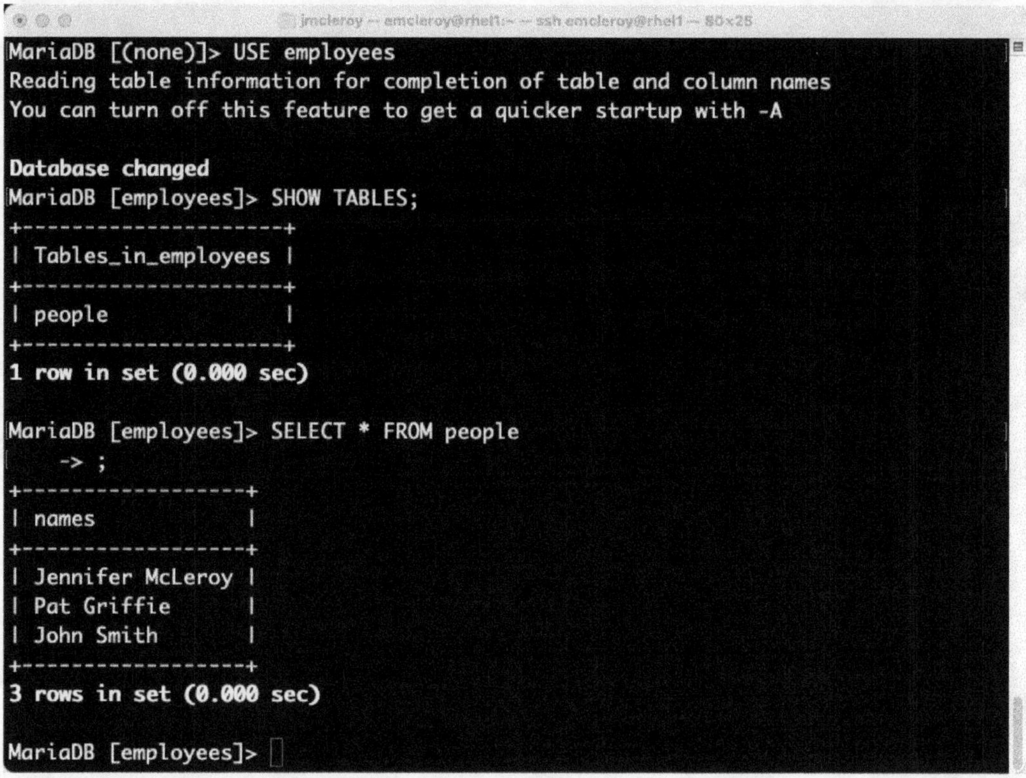

Figure 7.27 – Data is now populated within the system

In this section, we learned how to automate the installation and configuration of MariaDB. With this information, you can successfully deploy databases quickly and more confidently. The lower threshold for human error allows for greater confidence in the deployment and it can be done in seconds instead of hours, which is how long it would take somebody to do so with a keyboard for multiple databases. This is an advantage of using Ansible Automation.

Summary

In this chapter, we learned about the MariaDB SQL database and how it can be configured and updated. We learned how to automate it so that it can be easily deployed with the information we need it to have. This allows us to deploy a database quickly without the need for much human intervention. We also were able to back up the database and restore it, which is a crucial thing to be able to do when working with databases. In our next chapter, we will be talking about web servers, which use databases as their backends in some cases. We will go into installing and setting up these databases both manually and with Ansible Automation. I look forward to seeing you in the next chapter on our journey through the preparation for the EX358 exam.

8

Web Servers and Web Traffic – Learning How to Create and Control Traffic

In this chapter, we are going to overview what a web server is and install it. We will dive into securing the web server in order to ensure that the data is safe when you access it as a customer. We will also talk about how to automate this process using Ansible Automation. Without a good web server setup, you could possibly stunt your company's growth. The ability to provide access to your products is vital, and being able to safeguard your customer's information is equally as important. Let's go ahead and step into the world of web servers.

In this chapter, we're going to cover the following main topics:

- Getting started with web servers and traffic control
- Learning to set up web servers manually and control traffic
- Learning to use Ansible Automation to automate web servers and control traffic

Technical requirements

The technical requirements in this chapter are covered in the following sections.

Setting up GitHub access

Please refer to the instructions found in *Chapter 1, Block Storage – Learning How to Provision Block Storage on Red Hat Enterprise Linux*, to gain access to GitHub, and you will find the Ansible automation playbooks for this chapter at the following link: `https://github.com/PacktPublishing/Red-Hat-Certified-Specialist-in-Services-Management-and-Automation-EX358-Exam-Guide/tree/main/Chapter08`. Remember these are suggested playbooks and are not the only way you can write them to make the playbooks work for you.

You can always change them using raw, shell, or cmd to achieve the same results, but we are demonstrating the best way to accomplish our goals. Also, keep in mind that we are not using the FCQN needed in the future version of Ansible, as that will not be supported in the exam since it is testing against Ansible 2.9.

Getting started with web servers and traffic control

The ability to set up and run your own web server is important to most companies. In this digital age, you must be able to keep up with the demands of the fast-paced world. Without a strong website to offer your customers access to your product or services you will greatly hinder your sales potential. The ability to grow your business, especially internationally, is the dream of most people that start a new business. Setting up a web server on RHEL 8 is pretty straightforward using Apache httpd or NGINX but can become complex when implementing rules to protect yourself.

The ability to keep these same resources safe is a must for any business. The need to be able to block malicious traffic or spoofing is vital to keep your company in good standing with your customers. Being able to respond to threats from **distributed denial-of-service (DDoS)** attacks, for instance, is one of the things you need to be prepared for as a web server admin. We will go further, in future sections, into keeping your web server safe from unwanted access from third parties trying to gain your information for nefarious purposes.

We will showcase how to install a web server using Apache httpd and NGINX and how to configure it so that the system is secure. We will showcase how to house more than one web server endpoint on a physical server. Next, we will demonstrate how to access the web server once you have it up and running. We will do this using config files that will allow us to customize what is shown from the web server. Finally, we will automate all of this using Ansible Automation to take the headache out of deployments that need to be done over and over again.

Learning to set up web servers manually and control traffic

We are going to start with Apache httpd. We will commence by installing the service and configuring the files. We can see in the following screenshot the installation of Apache httpd:

```
Updating Subscription Management repositories.
Red Hat Enterprise Linux 8 for x86_64 - AppStre  11 kB/s | 4.5 kB     00:00
Red Hat Enterprise Linux 8 for x86_64 - BaseOS   12 kB/s | 4.1 kB     00:00
Dependencies resolved.
================================================================================
 Package              Arch   Version        Repository                     Size
================================================================================
Installing:
 httpd                x86_64 2.4.37-47.module+el8.6.0+15654+427eba2e.2
                                            rhel-8-for-x86_64-appstream-rpms 1.4 M
Installing dependencies:
 apr-util             x86_64 1.6.1-6.el8    rhel-8-for-x86_64-appstream-rpms 105 k
 apr                  x86_64 1.6.3-12.el8   rhel-8-for-x86_64-appstream-rpms 130 k
 mod_http2            x86_64 1.15.7-5.module+el8.6.0+13996+01710940
                                            rhel-8-for-x86_64-appstream-rpms 155 k
 httpd-tools          x86_64 2.4.37-47.module+el8.6.0+15654+427eba2e.2
                                            rhel-8-for-x86_64-appstream-rpms 108 k
 httpd-filesystem     noarch 2.4.37-47.module+el8.6.0+15654+427eba2e.2
                                            rhel-8-for-x86_64-appstream-rpms  41 k
 redhat-logos-httpd   noarch 84.5-1.el8     rhel-8-for-x86_64-baseos-rpms     29 k
Installing weak dependencies:
 apr-util-openssl     x86_64 1.6.1-6.el8    rhel-8-for-x86_64-appstream-rpms  27 k
 apr-util-bdb         x86_64 1.6.1-6.el8    rhel-8-for-x86_64-appstream-rpms  25 k
Enabling module streams:
 httpd                       2.4
```

Figure 8.1 – Installing Apache httpd web service

We are going to look at the default `httpd.conf` file in order to showcase the out-of-the-box setup for Apache httpd. We can see in the following screenshot the layout of the `/etc/httpd/conf/httpd.conf` file:

```
●  ●  ●                    jmcleroy — emcleroy@rhel1:~ — ssh emcleroy@rhel1 — 80×25
#
# This is the main Apache HTTP server configuration file.  It contains the
# configuration directives that give the server its instructions.
# See <URL:http://httpd.apache.org/docs/2.4/> for detailed information.
# In particular, see
# <URL:http://httpd.apache.org/docs/2.4/mod/directives.html>
# for a discussion of each configuration directive.
#
# See the httpd.conf(5) man page for more information on this configuration,
# and httpd.service(8) on using and configuring the httpd service.
#
# Do NOT simply read the instructions in here without understanding
# what they do.  They're here only as hints or reminders.  If you are unsure
# consult the online docs. You have been warned.
#
# Configuration and logfile names: If the filenames you specify for many
# of the server's control files begin with "/" (or "drive:/" for Win32), the
# server will use that explicit path.  If the filenames do *not* begin
# with "/", the value of ServerRoot is prepended -- so 'log/access_log'
# with ServerRoot set to '/www' will be interpreted by the
# server as '/www/log/access_log', where as '/log/access_log' will be
# interpreted as '/log/access_log'.
#
#
                                                            1,1              Top
```

Figure 8.2 – The /etc/httpd/conf/httpd.conf truncated file

As seen in the previous screenshot, information on how to set up a web server can be found in the Apache httpd **manual page** (**man**) pages. You can utilize these to understand what you are setting and to remind you of items you may have forgotten. The /etc/httpd/conf/httpd.conf file is also a good place to look for help and reminders, but do not rely on it alone.

Next, will use what we can learn from the man pages and the httpd.conf file to create a .conf file within the /etc/httpd/conf.d/ directory. This will allow us to create a web server to meet our needs. We are going to set up virtual hosts within Apache httpd and use the /etc/httpd/conf.d/ directory to store our websites. Before we do that, we have to ensure that the web server is set up within the default httpd.conf file, where we need to check the following code:

```
ServerRoot "/etc/httpd"
Listen 80
User apache
Group apache
```

The preceding code states that ServerRoot is /etc/httpd/, the web server Listen port is 80, and User for the web server is set to apache. These ensure that the ownership of the files is correct when serving the web server to the public and tell the server what port to listen on, which in this case would be port 80 or HTTP, which most of us are used to, or HTTPS which is more secure. For now, we will stick to this basic configuration.

Next, we will build the configuration for the virtual servers using the website.conf file, which we will save to /etc/httpd/conf.d/website.conf. We can see the contents of the website.conf file in the following screenshot, which displays the minimum required arguments to set up a web server:

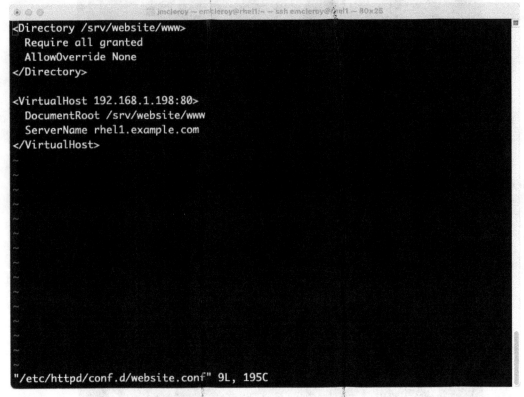

Figure 8.3 – The website.conf file used to configure the virtual web server

In the preceding screenshot, we can see that we are defining a directory for files, the IP of the server with a listen port of 80, and the website name. This will allow us to create files such as index.html in /srv/website/www in order to control the content of our web server. We need to create an index.html file and then start and enable the service. In the following screenshot, you can see that I have created a simple index.html file within /srv/website/www/:

```
Testing for EX358 book!!!




















-- INSERT --
```

Figure 8.4 – The index.hmtl file for Apache httpd web server

Next, we will start and enable httpd, as shown in the following screenshot:

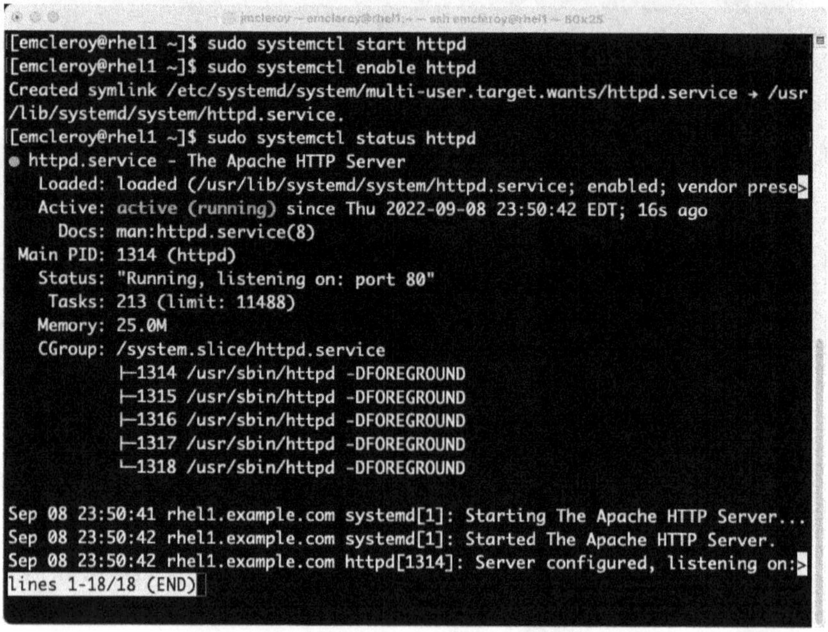

```
[emcleroy@rhel1 ~]$ sudo systemctl start httpd
[emcleroy@rhel1 ~]$ sudo systemctl enable httpd
Created symlink /etc/systemd/system/multi-user.target.wants/httpd.service → /usr
/lib/systemd/system/httpd.service.
[emcleroy@rhel1 ~]$ sudo systemctl status httpd
● httpd.service - The Apache HTTP Server
   Loaded: loaded (/usr/lib/systemd/system/httpd.service; enabled; vendor prese>
   Active: active (running) since Thu 2022-09-08 23:50:42 EDT; 16s ago
     Docs: man:httpd.service(8)
 Main PID: 1314 (httpd)
   Status: "Running, listening on: port 80"
    Tasks: 213 (limit: 11488)
   Memory: 25.0M
   CGroup: /system.slice/httpd.service
           ├─1314 /usr/sbin/httpd -DFOREGROUND
           ├─1315 /usr/sbin/httpd -DFOREGROUND
           ├─1316 /usr/sbin/httpd -DFOREGROUND
           ├─1317 /usr/sbin/httpd -DFOREGROUND
           └─1318 /usr/sbin/httpd -DFOREGROUND

Sep 08 23:50:41 rhel1.example.com systemd[1]: Starting The Apache HTTP Server...
Sep 08 23:50:42 rhel1.example.com systemd[1]: Started The Apache HTTP Server.
Sep 08 23:50:42 rhel1.example.com httpd[1314]: Server configured, listening on:>
lines 1-18/18 (END)
```

Figure 8.5 – Start, enable and status of Apache httpd

We now have to open the firewall rules. As a habit I suggest you adopt, we will open this for HTTP and HTTPS so that we can utilize both 80 and 443 if necessary to access our web server. In the following screenshot, we can see that we are opening the firewall rules, making them permanent, and then reloading the firewall:

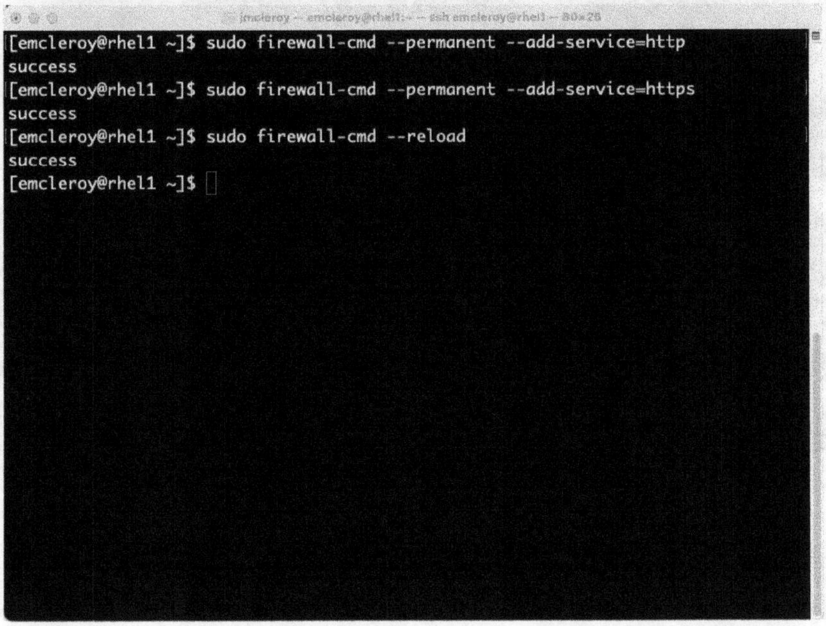

```
[emcleroy@rhel1 ~]$ sudo firewall-cmd --permanent --add-service=http
success
[emcleroy@rhel1 ~]$ sudo firewall-cmd --permanent --add-service=https
success
[emcleroy@rhel1 ~]$ sudo firewall-cmd --reload
success
[emcleroy@rhel1 ~]$ 
```

Figure 8.6 – Opening the firewall rules for Apache httpd

Next, we will restore the SELinux policy so that the web server showcases the correct information, which can be seen in the following screenshot:

```
[emcleroy@rhel1 ~]$ sudo restorecon -Rv /srv/
Relabeled /srv/website/www from unconfined_u:object_r:var_t:s0 to unconfined_u:object_r:httpd_sys_content_t:s0
Relabeled /srv/website/www/index.html from unconfined_u:object_r:var_t:s0 to unconfined_u:object_r:httpd_sys_content_t
:s0
[emcleroy@rhel1 ~]$
[emcleroy@rhel1 ~]$
[emcleroy@rhel1 ~]$
[emcleroy@rhel1 ~]$
[emcleroy@rhel1 ~]$
[emcleroy@rhel1 ~]$
[emcleroy@rhel1 ~]$
[emcleroy@rhel1 ~]$
[emcleroy@rhel1 ~]$
[emcleroy@rhel1 ~]$
[emcleroy@rhel1 ~]$
[emcleroy@rhel1 ~]$
[emcleroy@rhel1 ~]$
[emcleroy@rhel1 ~]$
[emcleroy@rhel1 ~]$
[emcleroy@rhel1 ~]$
[emcleroy@rhel1 ~]$
[emcleroy@rhel1 ~]$
[emcleroy@rhel1 ~]$
[emcleroy@rhel1 ~]$
[emcleroy@rhel1 ~]$
[emcleroy@rhel1 ~]$
[emcleroy@rhel1 ~]$
[emcleroy@rhel1 ~]$
```

Figure 8.7 – The SELinux command to restore the directory permissions

Finally, we are going to browse to the website, as shown in the following screenshot:

Testing for EX358 book!!!

Figure 8.8 – Successful web server browsing of the index.html file

Up until this point, we have successfully opened up our web server to the world. By having many servers behind a load balancer, you can alter where the traffic goes by the way the load balancer is configured.

Next, we will troubleshoot virtual hosts within Apache httpd. We will show you how to narrow down common problems when setting up virtual hosts and how to identify and fix them easily. First, let's create a virtual host with Apache httpd. One of the key things to note is the SELinux construct and how to restore it. The key takeaway from knowing this understanding what has caused the problem is if you receive the normal Apache httpd resource page, as seen in the following screenshot:

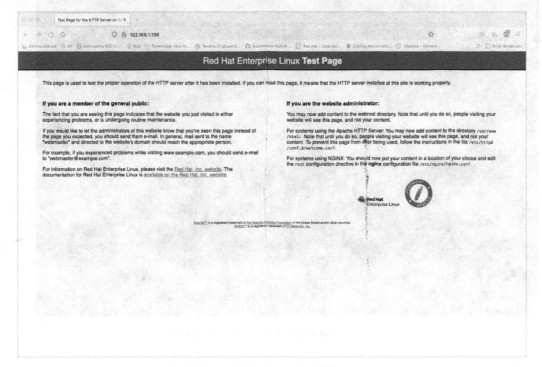

Figure 8.9 – The default Apache httpd page

After using the following command on the directory that you are using as the root directory for your server, you should see the **Testing for EX358book!!!** message:

```
[emcleroy@rhel1 html]$ sudo restorecon -rv /var/
```

The next thing you will want to do is look at the VirtualHost information found in /etc/httpd/conf.d/<name of the server>.conf for where the default files are held for the virtual host in case they are not in the default of /var/www/html. This way, you know where to restore the SELinux context in order to resolve the issue.

Next, make sure that you include the `Required all granted` command line for the directories; otherwise, you will not be able to see any as, by default, they are hidden from the world. This can be observed in the following virtual host screenshot:

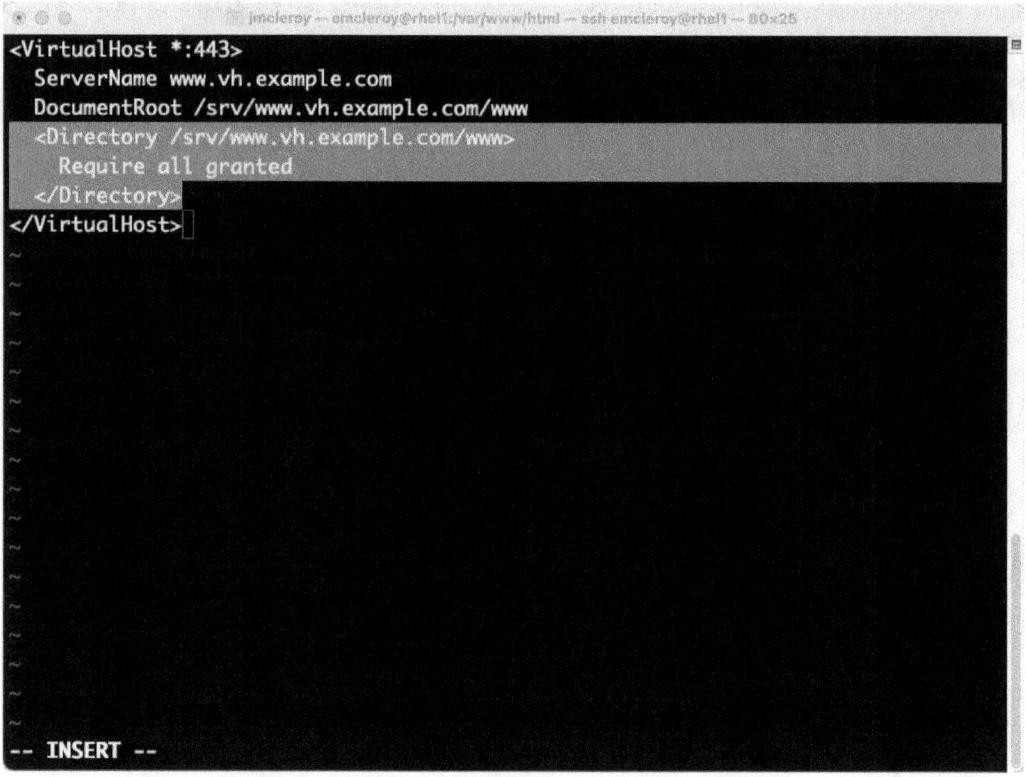

Figure 8.10 – Example of an httpd virtual host

We can use this information to know where to restore the SELinux context.

Finally, check the firewall rules to make sure they are allowing `443` in this instance, as shown in the following screenshot:

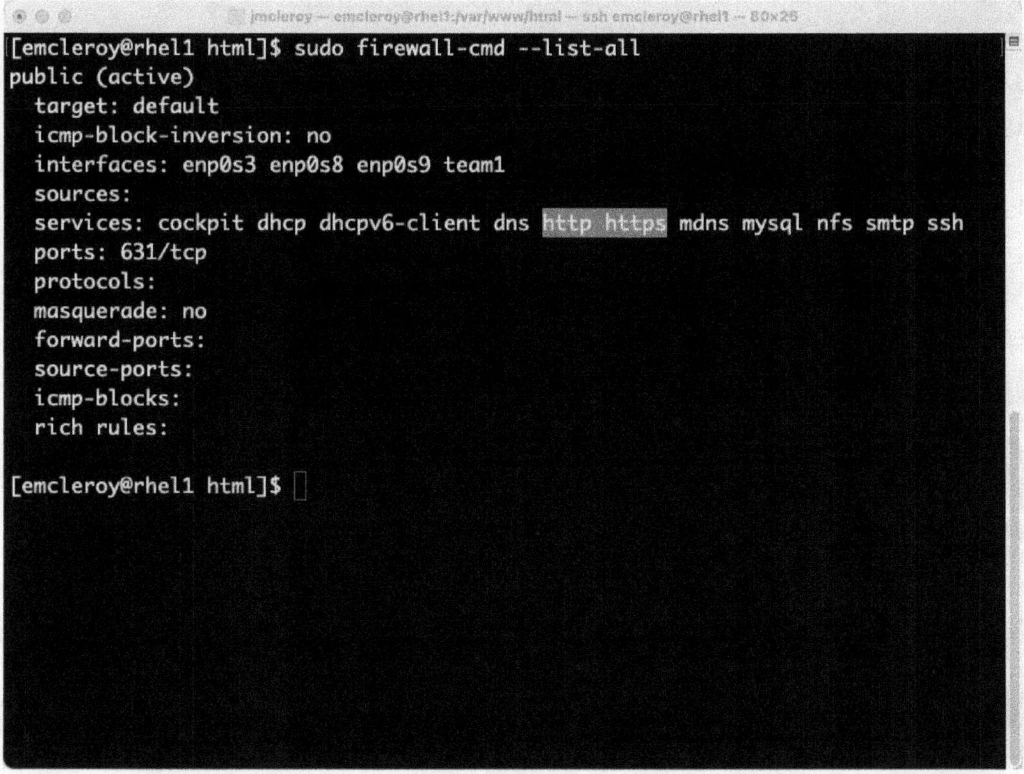

```
[emcleroy@rhel1 html]$ sudo firewall-cmd --list-all
public (active)
  target: default
  icmp-block-inversion: no
  interfaces: enp0s3 enp0s8 enp0s9 team1
  sources:
  services: cockpit dhcp dhcpv6-client dns http https mdns mysql nfs smtp ssh
  ports: 631/tcp
  protocols:
  masquerade: no
  forward-ports:
  source-ports:
  icmp-blocks:
  rich rules:

[emcleroy@rhel1 html]$
```

Figure 8.11 – Firewall rules show HTTP (80) and HTTPS (443) are both open

After we ensure the firewall rules are in place, the web server should function correctly. You can ensure that httpd is running through a status check, as shown in the following screenshot:

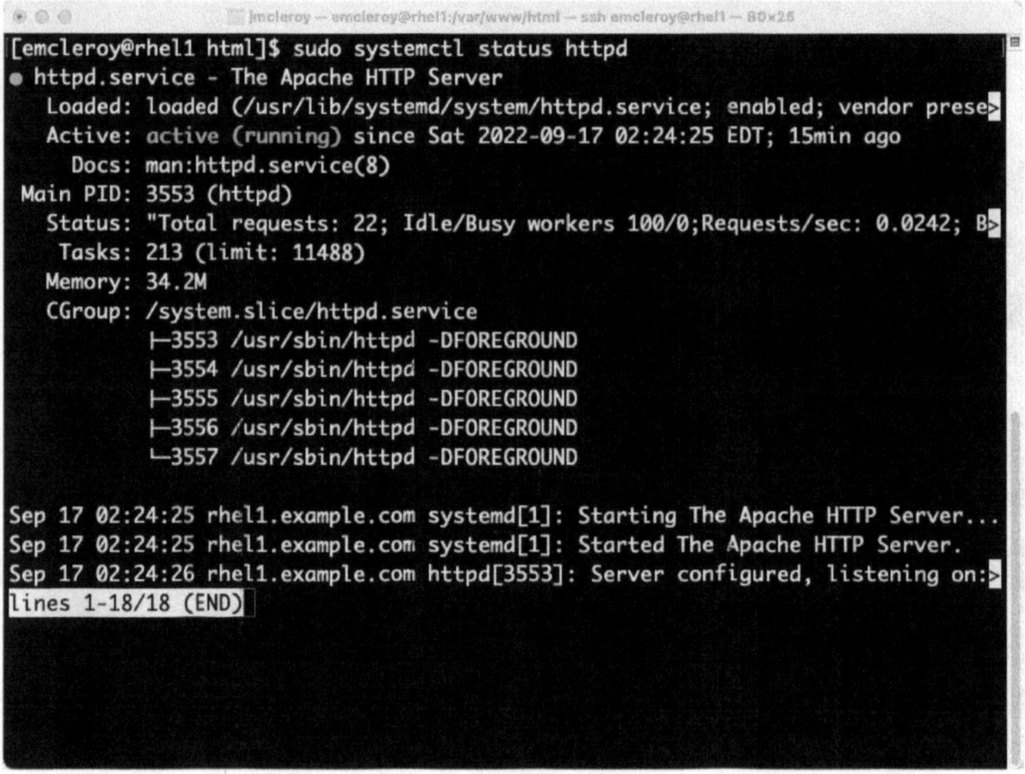

Figure 8.12 – Apache httpd status showing it is enabled and running

After the system status, firewall rules, and SELinux have been checked and corrected if needed, your web server should perform correctly.

Next, we will set up an Apache web server with TLS certificates. For this, we will use mod_ssl. We will start by installing mod_ssl using the following command:

```
[emcleroy@rhel1 ~]$ sudo dnf install mod_ssl -y
```

After you have installed it, you will need to modify the location of the certificates to match where you have them saved. This can be set in the /etc/httpd/conf.d/ssl.conf config file. The areas that need modification can be seen in the following screenshot:

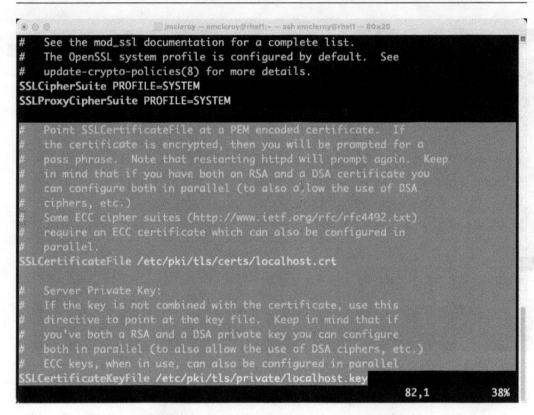

```
#  See the mod_ssl documentation for a complete list.
#  The OpenSSL system profile is configured by default.  See
#  update-crypto-policies(8) for more details.
SSLCipherSuite PROFILE=SYSTEM
SSLProxyCipherSuite PROFILE=SYSTEM


#  Point SSLCertificateFile at a PEM encoded certificate.  If
#  the certificate is encrypted, then you will be prompted for a
#  pass phrase.  Note that restarting httpd will prompt again.  Keep
#  in mind that if you have both an RSA and a DSA certificate you
#  can configure both in parallel (to also allow the use of DSA
#  ciphers, etc.)
#  Some ECC cipher suites (http://www.ietf.org/rfc/rfc4492.txt)
#  require an ECC certificate which can also be configured in
#  parallel.
SSLCertificateFile /etc/pki/tls/certs/localhost.crt


#  Server Private Key:
#  If the key is not combined with the certificate, use this
#  directive to point at the key file.  Keep in mind that if
#  you've both a RSA and a DSA private key you can configure
#  both in parallel (to also allow the use of DSA ciphers, etc.)
#  ECC keys, when in use, can also be configured in parallel
SSLCertificateKeyFile /etc/pki/tls/private/localhost.key
                                                    82,1          38%
```

Figure 8.13 – Key location for the mod_ssl HTTP web server

This will default the web server to TLS HTTPS.

We will now set up a web server using NGINX. This is a slightly different way of configuring web servers, and we will go into detail on how this is accomplished.

First, let's remove httpd and the firewall rules so that we have a fresh server to work on, which can be seen in the following screenshot:

```
 Erasing          : apr-util-bdb-1.6.1-6.el8.x86_64                                    7/9
 Erasing          : apr-util-openssl-1.6.1-6.el8.x86_64                                8/9
 Erasing          : mod_http2-1.15.7-5.module+el8.6.0+13996+01710940.x86_64            9/9
 Running scriptlet: mod_http2-1.15.7-5.module+el8.6.0+13996+01710940.x86_64            9/9
 Verifying        : apr-1.6.3-12.el8.x86_64                                            1/9
 Verifying        : apr-util-1.6.1-6.el8.x86_64                                        2/9
 Verifying        : apr-util-bdb-1.6.1-6.el8.x86_64                                    3/9
 Verifying        : apr-util-openssl-1.6.1-6.el8.x86_64                                4/9
 Verifying        : httpd-2.4.37-47.module+el8.6.0+15654+427eba2e.2.x86_64            5/9
 Verifying        : httpd-filesystem-2.4.37-47.module+el8.6.0+15654+427eba2e.2.noarch 6/9
 Verifying        : httpd-tools-2.4.37-47.module+el8.6.0+15654+427eba2e.2.x86_64      7/9
 Verifying        : mod_http2-1.15.7-5.module+el8.6.0+13996+01710940.x86_64           8/9
 Verifying        : redhat-logos-httpd-84.5-1.el8.noarch                              9/9
Installed products updated.

Removed:
  httpd-2.4.37-47.module+el8.6.0+15654+427eba2e.2.x86_64
  apr-1.6.3-12.el8.x86_64
  apr-util-1.6.1-6.el8.x86_64
  apr-util-bdb-1.6.1-6.el8.x86_64
  apr-util-openssl-1.6.1-6.el8.x86_64
  httpd-filesystem-2.4.37-47.module+el8.6.0+15654+427eba2e.2.noarch
  httpd-tools-2.4.37-47.module+el8.6.0+15654+427eba2e.2.x86_64
  mod_http2-1.15.7-5.module+el8.6.0+13996+01710940.x86_64
  redhat-logos-httpd-84.5-1.el8.noarch

Complete!
[emcleroy@rhel1 ~]$ sudo firewall-cmd --permanent --remove-service=http
success
[emcleroy@rhel1 ~]$ sudo firewall-cmd --permanent --remove-service=https
success
[emcleroy@rhel1 ~]$
```

Figure 8.14 – Removed Apache httpd and firewall rules

Now that Apache httpd has been removed, we can start with a fresh NGINX install. We will install NGINX using the commands in the following screenshot:

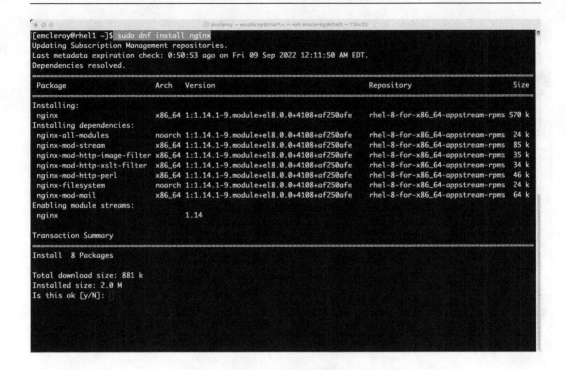

Figure 8.15 – Installing the NGINX web server

Next, we are going to configure the NGINX web server using the following files, which follow a similar pattern to that of Apache httpd. We will be looking at the main configuration file and server files separately. The main configuration file can be found at /etc/nginx/nginx.conf, and the server configuration files are found in the /etc/nginx/conf.d directory and have a .conf naming convention. The configuration of these files is slightly different from that of Apache httpd, as you can see in the following screenshot of the server configuration:

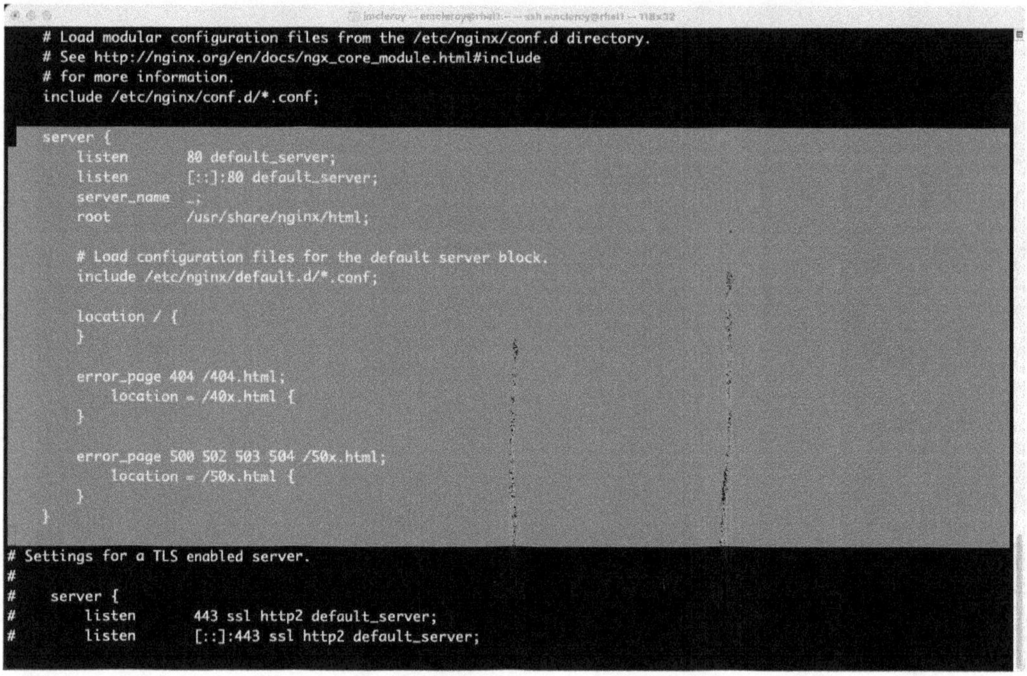

```
# Load modular configuration files from the /etc/nginx/conf.d directory.
# See http://nginx.org/en/docs/ngx_core_module.html#include
# for more information.
include /etc/nginx/conf.d/*.conf;

server {
    listen       80 default_server;
    listen       [::]:80 default_server;
    server_name  _;
    root         /usr/share/nginx/html;

    # Load configuration files for the default server block.
    include /etc/nginx/default.d/*.conf;

    location / {
    }

    error_page 404 /404.html;
        location = /40x.html {
    }

    error_page 500 502 503 504 /50x.html;
        location = /50x.html {
    }
}

# Settings for a TLS enabled server.
#
#    server {
#        listen       443 ssl http2 default_server;
#        listen       [::]:443 ssl http2 default_server;
```

Figure 8.16 – Configuration example of an NGINX web server

As you can see, the layout is different, and we will need to adjust it accordingly. We will now create a `website.conf` file and save it in the `/etc/nginx/conf.d` directory. An example of a web server setup for NGINX is shown in the following screenshot:

```
server {
    listen      80;
    server_name  rhel1.example.com;
    location / {
        root /var/www/html;
    }
}
```

```
"/etc/nginx/conf.d/website.conf" 7L, 147C
```

Figure 8.17 – Example of the configuration file for NGINX web server for rhel1

After we have saved the `website.conf` file, we will create `index.html` in the `/var/www/html` directory previously specified, as shown in the following screenshot:

```
Nginx for the WIN!!!
```

```
-- INSERT -- W10: Warning: Changing a readonly file                    1,18        All
```

Figure 8.18 – The index.html file for the NGINX web server

Following that, we will run the following command on the `/var/` directory in order to fix any SELinux issues:

```
[emcleroy@rhel1 ~]$ sudo restorecon -Rv /var/
Relabeled /var/log/dnf.librepo.log from system_u:object_r:rpm_
log_t:s0 to system_u:object_r:var_log_t:s0
Relabeled /var/log/dnf.rpm.log from system_u:object_r:rpm_
log_t:s0 to system_u:object_r:var_log_t:s0
Relabeled /var/log/Xorg.9.log.old from system_u:object_r:var_
log_t:s0 to system_u:object_r:xserver_log_t:s0
Relabeled /var/log/Xorg.9.log from system_u:object_r:var_
log_t:s0 to system_u:object_r:xserver_log_t:s0
Relabeled /var/log/dnf.log.1 from system_u:object_r:rpm_
log_t:s0 to system_u:object_r:var_log_t:s0
Relabeled /var/log/dnf.log from system_u:object_r:rpm_log_t:s0
to system_u:object_r:var_log_t:s0
```

Next, we will start, enable, and open the firewall rules as needed as if this was a fresh server, as shown in the following screenshot:

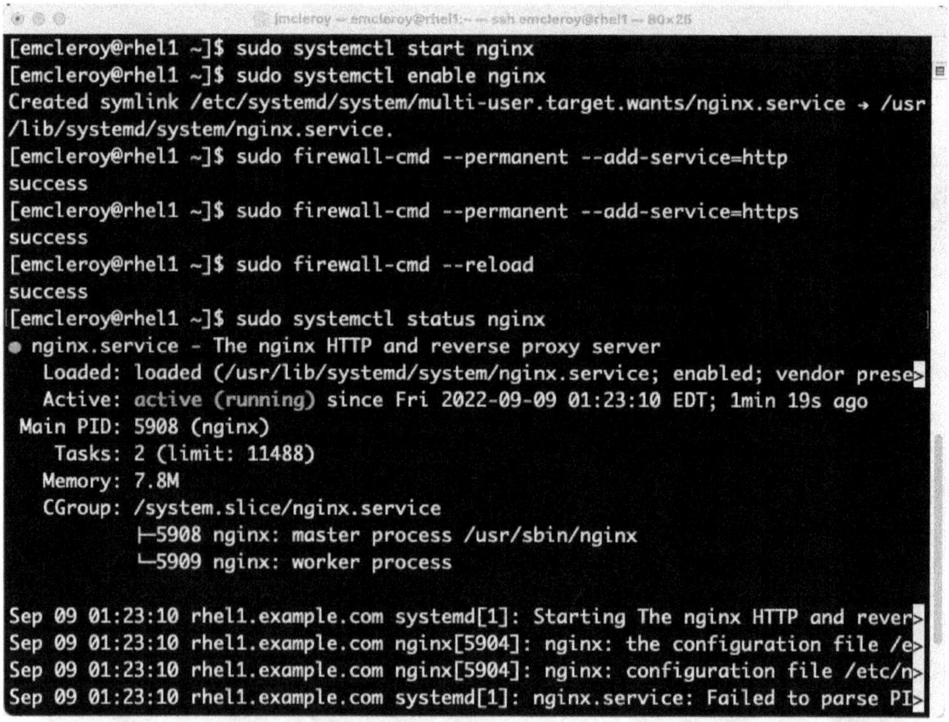

Figure 8.19 – Start, enable, and open firewall rules for the NGINX web server

Now, we can successfully browse to the new web server, as shown in the following screenshot:

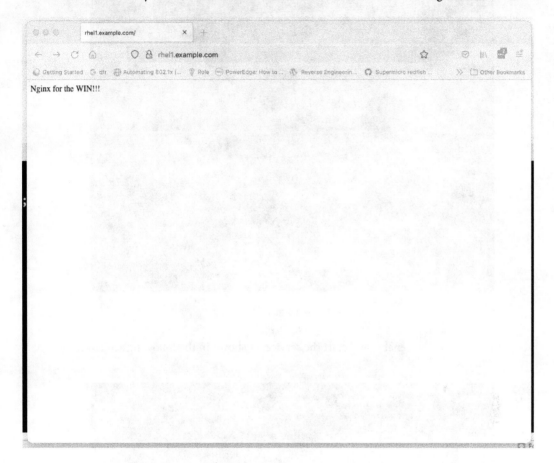

Figure 8.20 – Successful browsing to the NGINX web server

After we have our web servers set up, we are going to use HAProxy to control traffic to our servers. HAProxy provides load balancer abilities that enable us to determine what servers are currently in rotation for `roundrobin`, which allows connections to be dispersed on each new connection to a different web host for instance. By using HAProxy, we can lower our downtime during maintenance and ensure that we have high availability if a server fails.

We will start by installing HAProxy, as shown in the following screenshot:

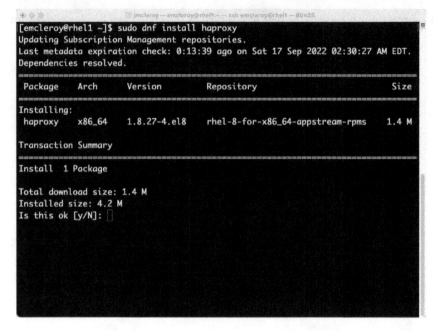

Figure 8.21 – Installing HAProxy

Once it is installed, we will enable and start the service, as shown in the following screenshot:

```
Downloading Packages:
haproxy-1.8.27-4.el8.x86_64.rpm                3.7 MB/s | 1.4 MB     00:00
--------------------------------------------------------------------------
Total                                          3.7 MB/s | 1.4 MB     00:00
Running transaction check
Transaction check succeeded.
Running transaction test
Transaction test succeeded.
Running transaction
  Preparing        :                                                    1/1
  Running scriptlet: haproxy-1.8.27-4.el8.x86_64                        1/1
  Installing       : haproxy-1.8.27-4.el8.x86_64                        1/1
  Running scriptlet: haproxy-1.8.27-4.el8.x86_64                        1/1
  Verifying        : haproxy-1.8.27-4.el8.x86_64                        1/1
Installed products updated.

Installed:
  haproxy-1.8.27-4.el8.x86_64

Complete!
[emcleroy@rhel1 ~]$ sudo systemctl start haproxy
[emcleroy@rhel1 ~]$ sudo systemctl enable haproxy
Created symlink /etc/systemd/system/multi-user.target.wants/haproxy.service → /u
sr/lib/systemd/system/haproxy.service.
[emcleroy@rhel1 ~]$
```

Figure 8.22 – Starting and enabling HAProxy

Once we have the service up and running, we will configure the main file that controls HAProxy, which is the `/etc/haproxy/haproxy.cfg` file, as shown in the following screenshot:

```
#---------------------------------------------------------------------
# main frontend which proxys to the backends
#---------------------------------------------------------------------
frontend main
    bind *:80
    acl url_static       path_beg      -i /static /images /javascript /styleshe
ets
    acl url_static       path_end      -i .jpg .gif .png .css .js

    use_backend static       if url_static
    default_backend          app

#---------------------------------------------------------------------
# static backend for serving up images, stylesheets and such
#---------------------------------------------------------------------
backend static
    balance      roundrobin
    server       static 127.0.0.1:4331 check

#---------------------------------------------------------------------
# round robin balancing between the various backends
#---------------------------------------------------------------------
backend app
    balance      roundrobin
-- INSERT --
```

Figure 8.23 – The /etc/haproxy/haproxy.cfg file

A takeaway from this review of the config file is that the listening port of the frontend has been changed to 80 as normal for HTTP traffic. The backend servers are where we should put the web servers we want to control access to. Finally, the round robin balancing is set to `roundrobin`, which will rotate whichever backend server the user hits when attempting to browse to the web server.

One thing to take away from HAProxy is that by default, SELinux allows many ports to be utilized out of the box, such as 80 and 443. However, if you wanted to change to something that was not a well-known HTTP or HTTPS port, you would need to make sure that the following command was set to allow other ports for HAProxy through SELinux:

```
[emcleroy@rhel1 ~]$ sudo setsebool -P haproxy_connect_any on
```

In this section, we learned how to set up Apache httpd, NGINX, and how to use HAProxy as a load balancer. Next, we will work on automating these sections to make life easier regardless of what you choose to deploy, whether Apache or NGINX, with HAProxy able to utilize both as backend servers.

Learning how to use Ansible Automation to automate web servers and control traffic

We will set up a multi-task playbook that allows Apache httpd to be installed on the server rhel1 and NGINX to be installed on rhel2. After that, we will install HAProxy and set the two servers to roundrobin load balance, all through the use of Ansible Automation.

Let's start with the inventory, which will include the httpd web server, NGINX web server, and the HAProxy server, as shown in the following screenshot:

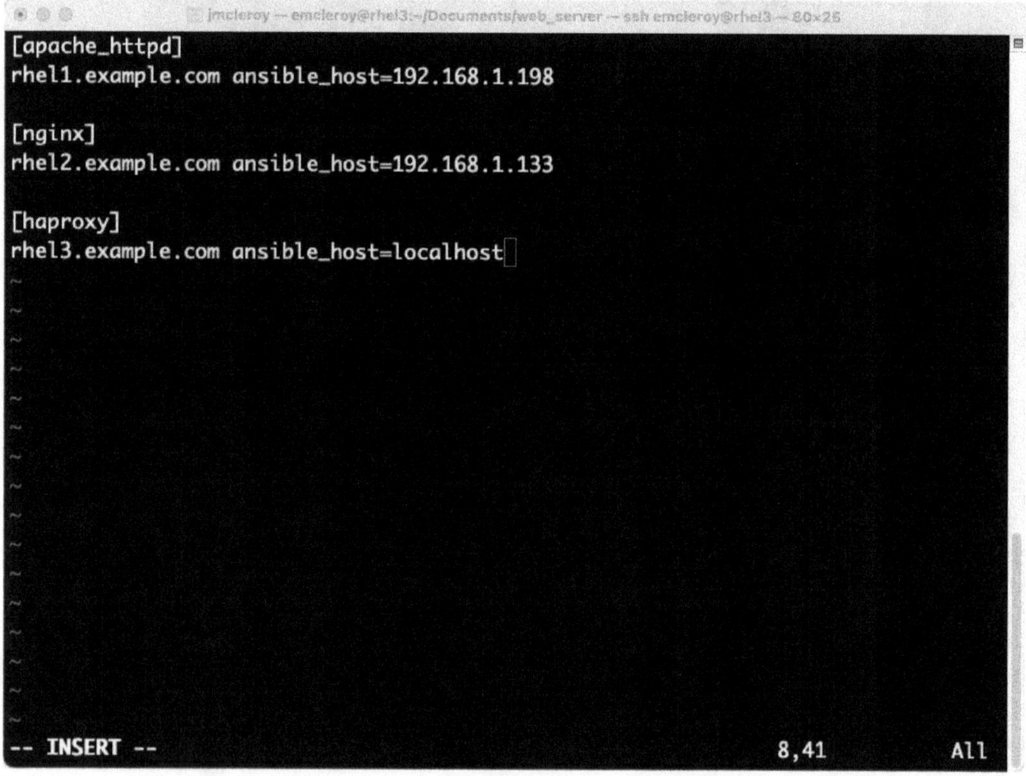

```
[apache_httpd]
rhel1.example.com ansible_host=192.168.1.198

[nginx]
rhel2.example.com ansible_host=192.168.1.133

[haproxy]
rhel3.example.com ansible_host=localhost
```

Figure 8.24 – Ansible Automation inventory file

Next, we will write out the playbook and files needed to make this work successfully. To do that, first, we will write out the beginning part of the playbook for the first server shown as follows:

```
---

- name: Install and configure Apache httpd
  hosts: rhel1.example.com
  become: true
```

```
become_method: sudo
tasks:
  - name: Install Apache httpd
    package:
      name: httpd
      state: latest
  - name: Copy Apache httpd web server configuration
    template:
      src: httpd.j2
      dest: /etc/httpd/conf.d/website.conf
  - name: Copy Apache httpd web server index.html
    template:
      src: httpd_html.j2
      dest: /var/www/html/index.html
  - name: Restore SELinux
    command:
      cmd: restorecon -rv /var/
  - name: Add firewall rules
    firewalld:
      service: http
      state: enabled
      permanent: true
  - name: Reload firewall
    command:
      cmd: firewall-cmd --reload
  - name: Start and enable Apache httpd
    service:
      name: httpd
      state: started
      enabled: true
```

The files mentioned for copying are displayed in the following screenshot:

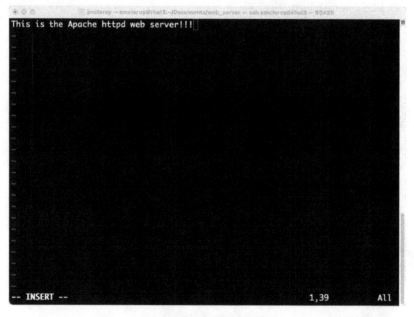

```
<Directory /var/html/www>
    Require all granted
    AllowOverride None
</Directory>

<VirtualHost 192.168.1.198:80>
    DocumentRoot /var/html/www/
    ServerName rhel1.example.com
</VirtualHost>




-- INSERT --                                                9,15        All
```

Figure 8.25 – httpd.j2

In the following screenshot, we can see the template .j2 file for the httpd_html conversion when we configure the web host:

```
This is the Apache httpd web server!!!




-- INSERT --                                                1,39        All
```

Figure 8.26 – httpd_html.j2

Next, we will configure the NGINX web server within the same playbook, as seen in the following code:

```
        state: started
        enabled: true


- name: Install and configure nginx
  hosts: rhel2.example.com
  become: true
  become_method: sudo
  tasks:
    - name: Install Nginx
      package:
        name: nginx
        state: latest
    - name: Create web server directory
      file:
        path: /srv/website/www
        state: directory
        mode: '0755'
    - name: Copy Nginx web server configuration
      template:
        src: nginx.j2
        dest: /etc/nginx/conf.d/website.conf
    - name: Copy Nginx web server index.html
      template:
        src: nginx_html.j2
        dest: /srv/website/www/index.html
    - name: Restore SELinux
      command:
        cmd: restorecon -rv /srv/
    - name: Add firewall rules
      firewalld:
        service: http
        state: enabled
        permanent: true
    - name: Reload firewall
      command:
```

```
        cmd: firewall-cmd --reload
  - name: Start and enable Nginx
    service:
      name: nginx
      state: started
      enabled: true
```

The files used within this portion of the playbook can be found in the following screenshot:

```
server {
    listen      80;
    server_name  rhel2.example.com;
    location / {
        root /srv/website/www;
    }
}
```

-- INSERT --   5,34   All

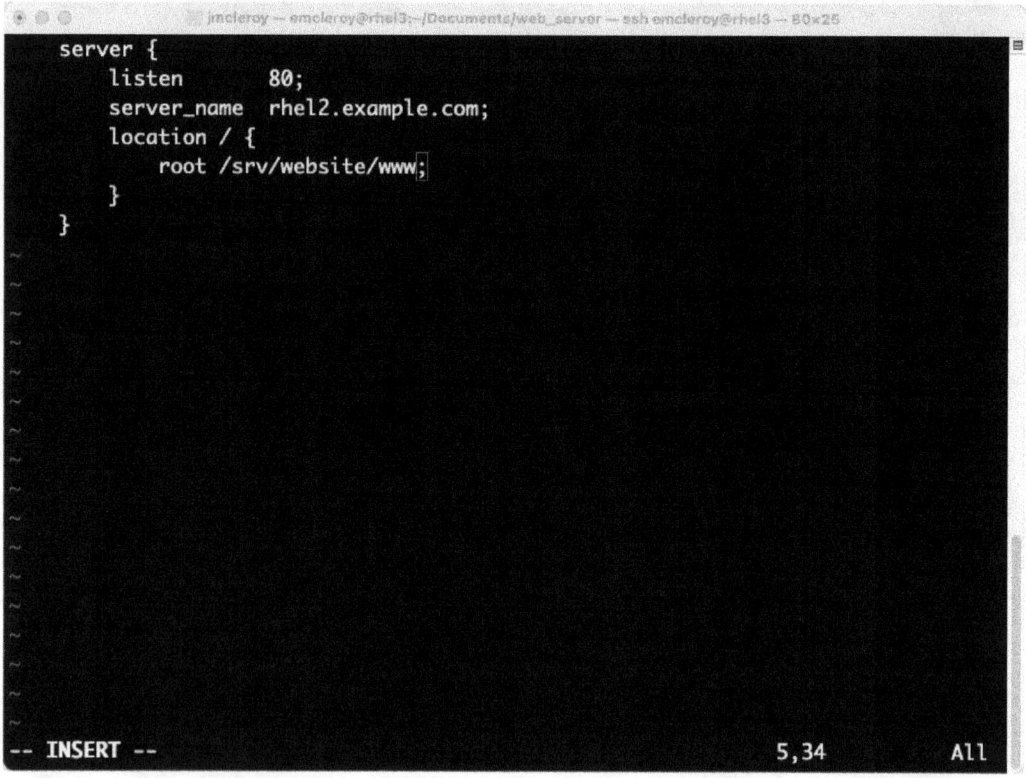

Figure 8.27 – nginx.j2

In the following screenshot, we can see the .j2 file that allows us to copy over the configuration of the NGINX web server:

```
#--------------------------------------------------------------------
frontend main
    bind *:80
    acl url_static          path_beg         -i /static /images /javascript /styleshe
ets
    acl url_static          path_end         -i .jpg .gif .png .css .js

    use_backend static           if url_static
    default_backend              app


#--------------------------------------------------------------------
# static backend for serving up images, stylesheets and such
#--------------------------------------------------------------------
backend static
    balance        roundrobin
    server         static 127.0.0.1:4331 check


#--------------------------------------------------------------------
# round robin balancing between the various backends
#--------------------------------------------------------------------
backend app
    balance        roundrobin
    server  app1 192.168.1.198:80 check
    server  app2 192.168.1.133:80 check
-- INSERT --                                              76,40           Bot
```

Figure 8.28 – nginx_html.j2

After those web servers are configured, we will need to set up HAProxy, as shown in the remainder of the playbook:

```
    - name: Start and enable Nginx
      service:
        name: nginx
        state: started
        enabled: true


  - name: Install and configure haproxy
    hosts: rhel3.example.com
    become: true
    become_method: sudo
```

```
tasks:
  - name: Install haproxy
    package:
      name: haproxy
      state: latest
  - name: Copy configuration file for haproxy
    template:
      src: haproxy.j2
      dest: /etc/haproxy/haproxy.cfg
  - name: Add firewall rules
    firewalld:
      service: http
      state: enabled
      permanent: true
  - name: Reload firewall
    command:
      cmd: firewall-cmd --reload
  - name: Start and enable haproxy
    service:
      name: haproxy
      state: started
      enabled: true
```

The file used to configure the HAProxy can be found in the following screenshot:

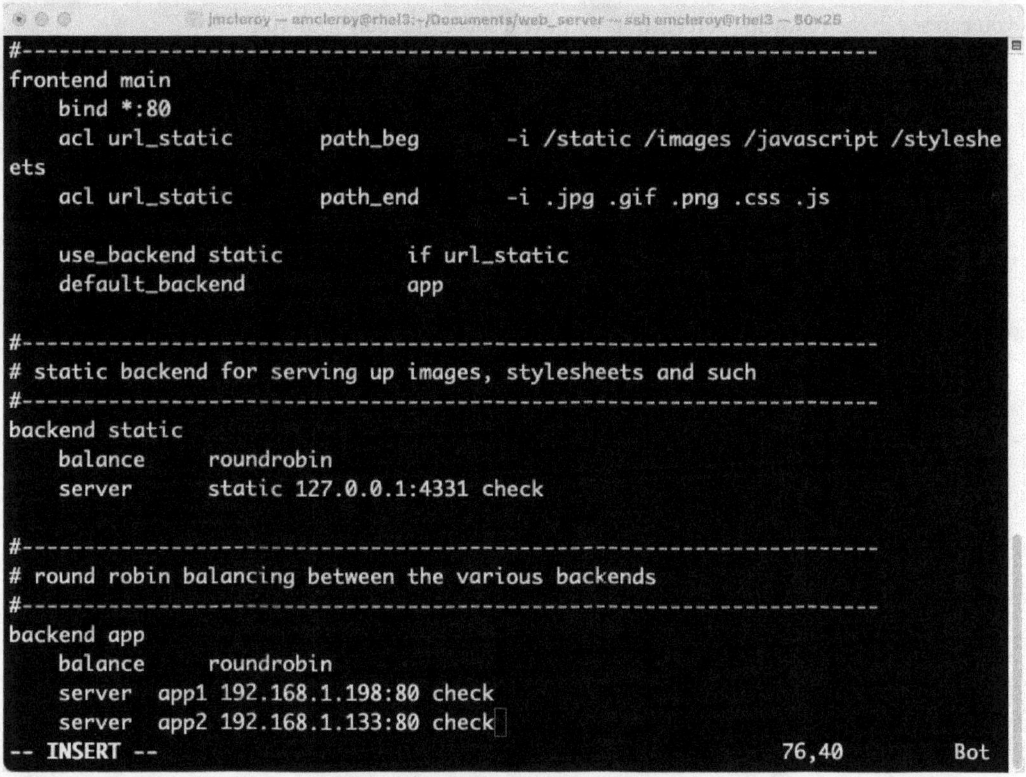

```
#---------------------------------------------------------------------
frontend main
    bind *:80
    acl url_static          path_beg        -i /static /images /javascript /styleshe
ets
    acl url_static          path_end        -i .jpg .gif .png .css .js

    use_backend static          if url_static
    default_backend             app


#---------------------------------------------------------------------
# static backend for serving up images, stylesheets and such
#---------------------------------------------------------------------
backend static
    balance     roundrobin
    server      static 127.0.0.1:4331 check


#---------------------------------------------------------------------
# round robin balancing between the various backends
#---------------------------------------------------------------------
backend app
    balance     roundrobin
    server  app1 192.168.1.198:80 check
    server  app2 192.168.1.133:80 check
-- INSERT --                                              76,40           Bot
```

Figure 8.29 – haproxy.j2

After we have created the playbook and the supporting files, we run the playbook as follows:

```
ansible-playbook -i inventory web_servers.yml -u emcleroy -k
--ask-become
```

As you can see in the following screenshot, we have a successful playbook run:

```
TASK [Install haproxy] ****************************************************
changed: [rhel3.example.com]

TASK [Copy configuration file for haproxy] ********************************
changed: [rhel3.example.com]

TASK [Add firewall rules] *************************************************
changed: [rhel3.example.com]

TASK [Reload firewall] ****************************************************
changed: [rhel3.example.com]

TASK [Start and enable haproxy] *******************************************
changed: [rhel3.example.com]

PLAY RECAP ****************************************************************
rhel1.example.com          : ok=8    changed=2    unreachable=0    failed=0    s
kipped=0    rescued=0    ignored=0
rhel2.example.com          : ok=9    changed=2    unreachable=0    failed=0    s
kipped=0    rescued=0    ignored=0
rhel3.example.com          : ok=6    changed=5    unreachable=0    failed=0    s
kipped=0    rescued=0    ignored=0

[emcleroy@rhel3 web_server]$
```

Figure 8.30 – Successful Ansible Automation playbook run

After the playbook runs successfully, we can now browse to rhel3.example.com, and it will round robin through the two backend servers.

In this section, we learned how to create both Apache httpd web servers and NGINX web servers. We also learned how to set up HAProxy to load balance and provide high availability to our web servers. This enables us to take servers out of rotation without our customers knowing what is happening, all through Ansible Automation playbooks.

Summary

In this chapter, we learned about web servers and how we can create different flavors from Apache to NGINX. We learned how to control server access and to resolve issues setting them up. Through HAProxy, we learned how to stir traffic through load balancing in order to ensure that we have servers up and running for the end users at all times. In the next chapter, we will wrap up what we have learned throughout this book with an overview followed by practice exam questions. This will help prepare you and drive home the information you have learned while studying this book. In the next chapter, we will work with mock exam questions to prepare you for test day. I look forward to seeing you all there and helping you succeed in passing the EX358 certification exam.

## 9

Comprehensive Review and Test Exam Questions

In this chapter, we will go over the different things we have learned throughout this book. We will be working through mock exam questions that will help familiarize you with the format of questions that you may encounter on the exam. We will attempt to cover all things that could possibly be on the exam per the public listed information by Red Hat, which can be found here: https://www. redhat.com/en/services/training/ex358-red-hat-certified-specialist-services-management-automation-exam?section=Objectives.

The outcome of this chapter will be preparing you fully to take the exam confidently and give you the information needed for you to pass the exam. We will go into other things that may not have been touched on in the book until now but will explain each topic as we go. We will also be covering the request to complete each question manually and to also do so with Ansible Automation. Let's get started.

In this chapter, we're going to cover the following:

- A comprehensive review of all exam objectives and mock exams for you to test your newfound skills

Technical requirements

We will be using the setup of the systems that is used in *Chapter 1, Block Storage – Learning How to Provision Block Storage on Red Hat Enterprise Linux*. We will utilize this setup along with the setup from the additional **Network Interface Controllers** (**NICs**) for the teaming exercise, which can be found in *Chapter 4, Link Aggregation Creation – Creating Your Own Link and Mastering the Networking Domain*. This will allow you to utilize all the upcoming questions as if you were in a test environment.

Setting up GitHub access

Please refer to the instructions found in *Chapter 1, Block Storage – Learning How to Provision Block Storage on Red Hat Enterprise Linux*, to gain access to GitHub for the book's repo. You will find the

Ansible Automation playbooks for this chapter at the following link: `https://github.com/PacktPublishing/Red-Hat-Certified-Specialist-in-Services-Management-and-Automation-EX358-Exam-Guide/tree/main/Chapter09`. Remember these are suggested playbooks and are not the only way you can write them to make them work for you.

You can always change them up using `raw`, `shell`, or `cmd` to achieve the same results, but we are demonstrating the best way to accomplish the goals. Also, keep in mind that we are not using the FCQN that is needed in future versions of Ansible, as that will not be supported in the exam, as it tests against Ansible 2.9.

A comprehensive review of all exam objectives and mock exams for you to test your newfound skills

We will present a mock exam with a mix of questions based on the skills you have learned throughout the entire book. We will go into more detail based on the items that you might run into in the exam itself. This will help prepare you to take the EX358 exam. These are not official questions; otherwise, they would not be provided in this book. Instead, they are theoretical scenarios you may face with differing information. Please set up your configuration as noted at the beginning of the chapter and then take note of your IP addresses, system names, and so on. This will allow you to grab the needed information fast for use within the exam so that you are not slowed down, as you only have 4 hours.

Let's get started with the exam. We will start with the information for my lab to give you an idea of how to write this out for your setup:

```
Username: emcleroy
Password: redhat
Admin username: root
Admin password: redhat
Node1: rhel1.example.com 192.168.1.198
Node2: rhel2.example.com 192.168.1.133
Node3: rhel3.example.com 192.168.1.53    Ansible control node
```

You can see that I have access to the information I need. If you set up your configuration to use hostnames through either DNS or your host file, you can connect to your devices easily. This also makes using the inventory a little easier in the Ansible inventory, as you do not have to add the `ansible_host` parameter. For our purposes, I will still be using `ansible_host` in my inventory file.

Another thing that makes everything easier is setting up RSA keys and sharing them across the system instead of using passwords. This is shown in the *Chapter 1, Block Storage – Learning How to Provision Block Storage on Red Hat Enterprise Linux*, setup process as well. This makes running the playbooks easier, as they do not need to be run with the `-u emcleroy -k --ask-become` portion of the command added to pass the SSH and the escalated privilege password. We will be using the command with the extra parameters to make sure that you understand how to use the full command, if necessary, during the exam.

After you have set up your system as required, we will get started with the mock exam and go through each topic to review the items we have learned about. The questions will be in the order laid out by Red Hat on the objectives page found here: `https://www.redhat.com/en/services/ training/ex358-red-hat-certified-specialist-services-management- automation-exam?section=Objectives`.

The questions follow. The answers will come directly after the questions, but please try to answer the questions without looking at the answers. I recommend attempting the questions once by yourself before looking at the solution. If you get feel stuck, feel free to read the answers and go through the solution to refresh your memory.

Managing network services

Managing network services is our first objective as noted by Red Hat for this exam.

First up, we will set up IP connectivity for the network interfaces.

Question 1 – For this question, we would like you to provision the NICs on the servers with static addresses. Using the DHCP IP addresses, we would like you to change to the static IPs that were noted at the start of the instructions. (These will differ depending on your network):

```
Rhel1 DHCP IP: 192.168.1.81
Rhel2 DHCP IP: 192.168.1.65

Rhel1 Static IP: 192.168.1.198
Rhel2 Static IP: 192.168.1.133
The subnet is 192.168.1.0/24
```

Once these are set up statically, you should be able to route between the two.

Answer 1 – To set a static IP up manually, we will need to SSH to the device using the DHCP IP addresses and then modify them to the static addresses.

First, we will use the `nmtui` command:

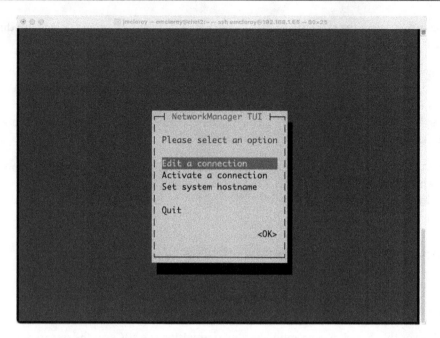

Figure 9.1 – NetworkManager TUI initial display

We will then choose to edit the connection and, from there, you will see the interfaces available, as shown in the following screenshot:

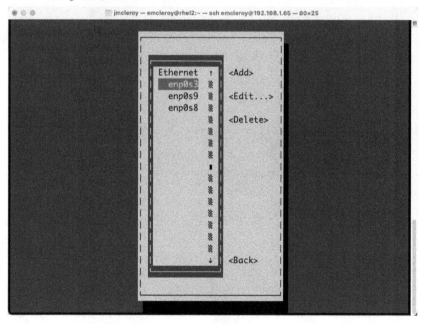

Figure 9.2 – NMTUI interface choices

After choosing which interface to change, we will then move on to the settings to enable a static IP, as shown in the following screenshot:

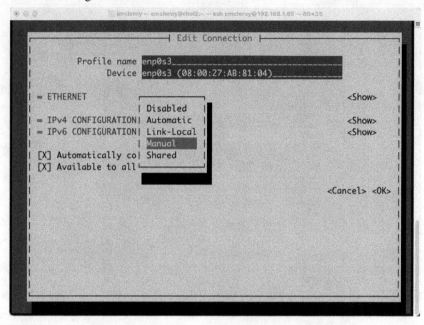

Figure 9.3 – Manual setting choice for the selected interface

We will then configure the required settings, such as the static IP, gateway, and DNS server, as shown in the following screenshot:

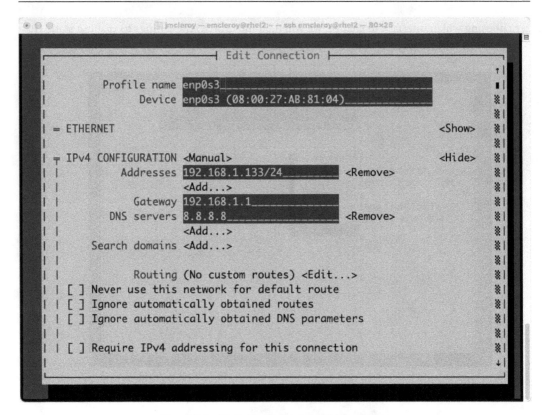

Figure 9.4 – Static IP, gateway, and DNS settings

After we have saved these settings, we will need to restart the network interface in order for it to change to the newly configured IP. We will do this by using the following command:

```
[emcleroy@rhel1 ~]$ sudo systemctl restart NetworkManager
```

Next, we will test to ensure the new address is reachable, as shown using the `ping` test in the following screenshot:

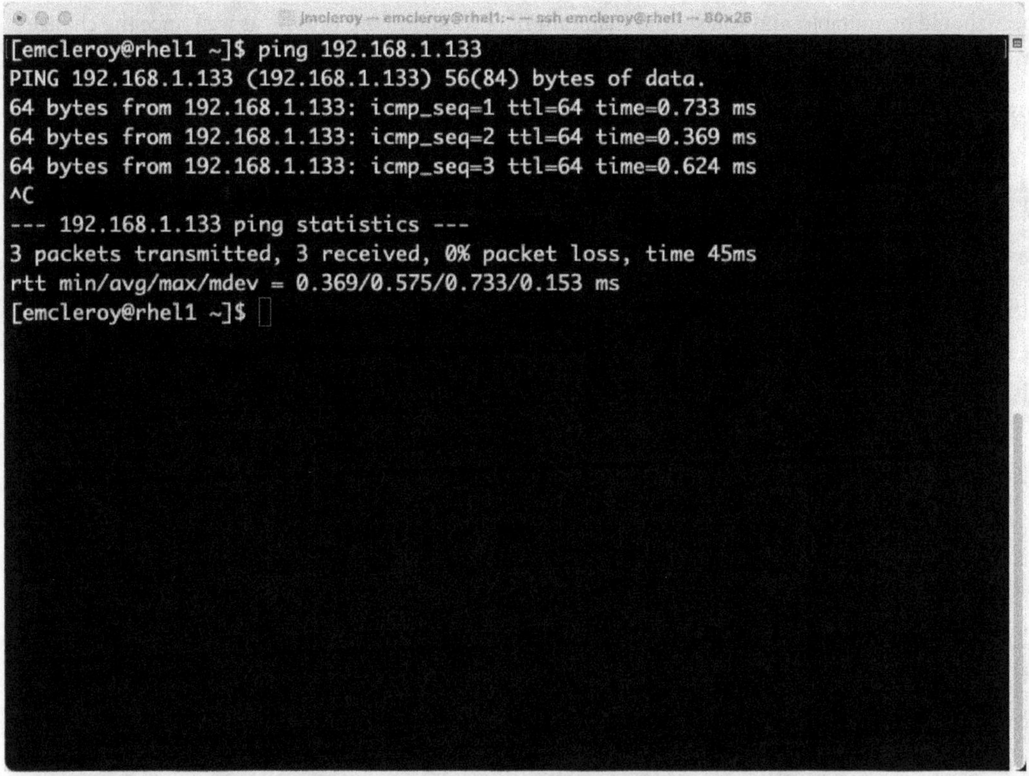
jmcleroy — emcleroy@rhel1:~ — ssh emcleroy@rhel1 — 80x28
[emcleroy@rhel1 ~]$ ping 192.168.1.133
PING 192.168.1.133 (192.168.1.133) 56(84) bytes of data.
64 bytes from 192.168.1.133: icmp_seq=1 ttl=64 time=0.733 ms
64 bytes from 192.168.1.133: icmp_seq=2 ttl=64 time=0.369 ms
64 bytes from 192.168.1.133: icmp_seq=3 ttl=64 time=0.624 ms
^C
--- 192.168.1.133 ping statistics ---
3 packets transmitted, 3 received, 0% packet loss, time 45ms
rtt min/avg/max/mdev = 0.369/0.575/0.733/0.153 ms
[emcleroy@rhel1 ~]$

Figure 9.5 – IP ping test to ensure connectivity

Next up, we will dig into setting up the IPV6 address configuration for previously configured network interfaces.

Question 2 – For this question, we would like you to set up an IPv6 address on the already-configured NIC while maintaining connectivity via IPv4:

```
Rhel1 IPv6: fc65:8956:7254:6321::a7/64
Rhel2 IPv6: fc65:8956:7254:6321::a8/64
Subnet: fc65:8956:7254:6321::/64
```

These should be routable, and you should maintain IPv4 routable connectivity as well once the change has been made.

Answer 2 – We will use the `nmtui` command to set this up:

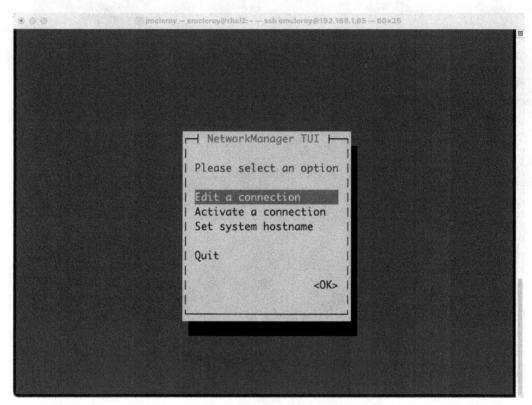

Figure 9.6 – NMTUI network menu

After we initiate the `nmtui` menu, we will then choose to edit a connection, which will take you to the menu in the following screenshot:

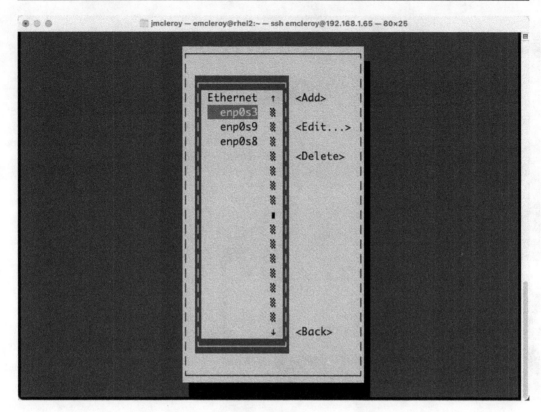

Figure 9.7 – NMTUI interface selection screen

After choosing the correct interface – in our case, **enp0s3**, but in your setup, it may be different depending on your lab – we will then be able to make configuration changes, as shown in the following screenshot:

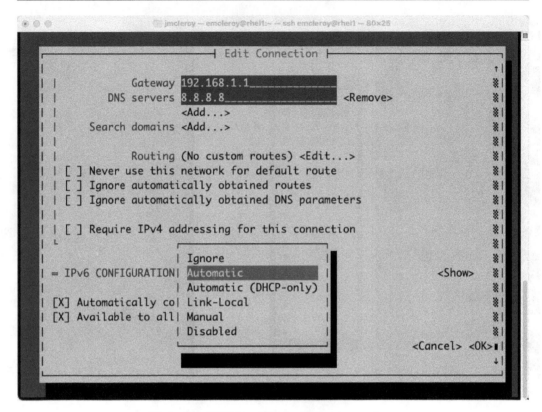

Figure 9.8 – NMTUI interface configuration settings

After we choose **Manual** for **IPv6 CONFIGURATION**, we can then edit the values of the IPv6 address and gateway, as shown in the following screenshot:

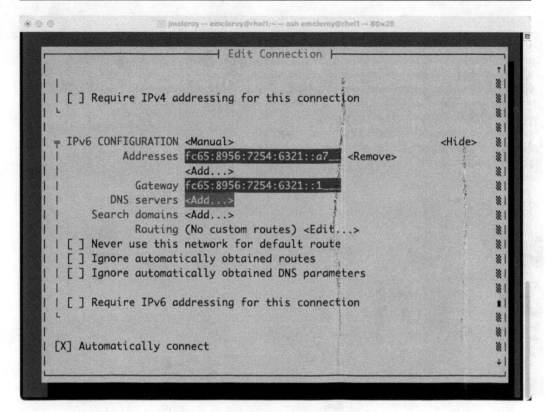

Figure 9.9 – IPv6 interface settings

After we set up the configuration as shown, we will then save the configuration. In order for the changes to take effect, we will need to run the following command to refresh the connectivity:

```
[emcleroy@rhel1 ~]$ sudo systemctl restart NetworkManager
```

After both servers have been configured, we can then run IPv6 `ping` tests across the two interfaces to ensure connectivity, as shown in the following screenshot:

```
[emcleroy@rhel1 ~]$ ping6 fc65:8956:7254:6321::a8
PING fc65:8956:7254:6321::a8(fc65:8956:7254:6321::a8) 56 data bytes
64 bytes from fc65:8956:7254:6321::a8: icmp_seq=1 ttl=64 time=0.507 ms
64 bytes from fc65:8956:7254:6321::a8: icmp_seq=2 ttl=64 time=0.292 ms
64 bytes from fc65:8956:7254:6321::a8: icmp_seq=3 ttl=64 time=0.433 ms
^C
--- fc65:8956:7254:6321::a8 ping statistics ---
3 packets transmitted, 3 received, 0% packet loss, time 35ms
rtt min/avg/max/mdev = 0.292/0.410/0.507/0.092 ms
[emcleroy@rhel1 ~]$
```

Figure 9.10 – IPv6 ping tests

We have shown one way to solve the issue of configuring your network with the preceding information. Now, we will move on to working with firewall services.

Managing firewall services

In this section, we will go over setting up firewall rules. We will talk about rich rules and other items that will block specific connectivity and allow connectivity for network traffic that is not currently allowed.

Question 1 – For this question, we would like you to allow access to services and ports through the firewall. This will allow connectivity to web services that might be running on your servers in production. Please open access to the following:

```
Services: http and https
```

Please do this manually and also set it up in an Ansible Automation playbook.

Answer 1 – We will use the following commands to set up the firewall to allow connectivity to these services:

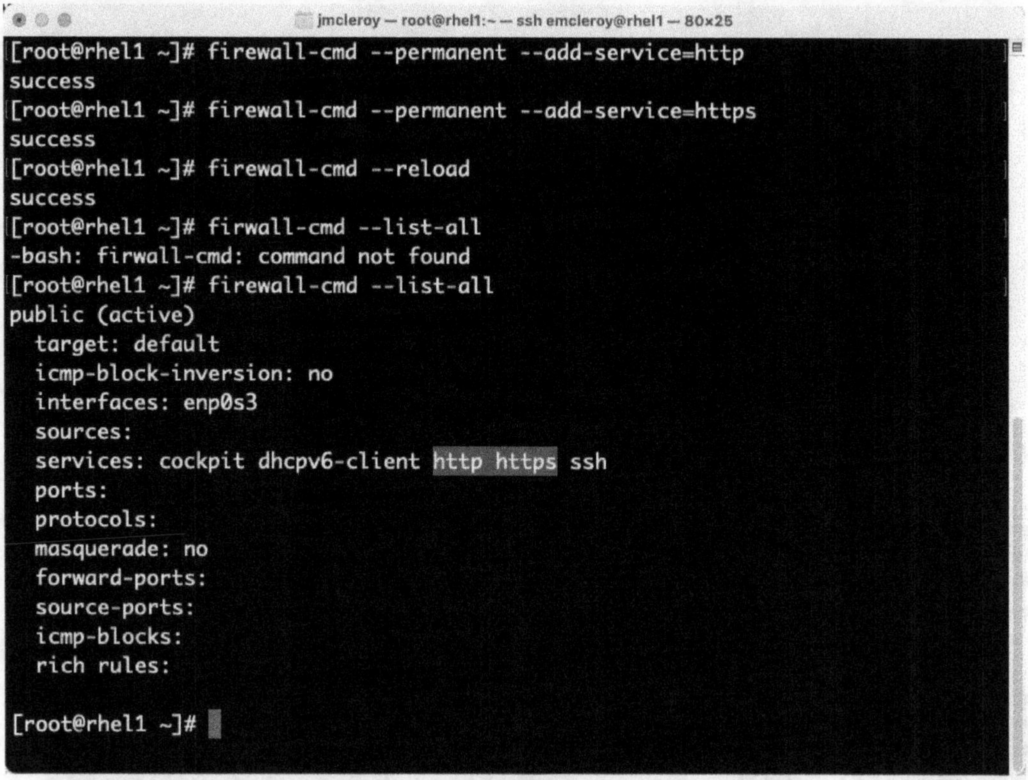

```
jmcleroy — root@rhel1:~ — ssh emcleroy@rhel1 — 80×25
[root@rhel1 ~]# firewall-cmd --permanent --add-service=http
success
[root@rhel1 ~]# firewall-cmd --permanent --add-service=https
success
[root@rhel1 ~]# firewall-cmd --reload
success
[root@rhel1 ~]# firwall-cmd --list-all
-bash: firwall-cmd: command not found
[root@rhel1 ~]# firewall-cmd --list-all
public (active)
  target: default
  icmp-block-inversion: no
  interfaces: enp0s3
  sources:
  services: cockpit dhcpv6-client http https ssh
  ports:
  protocols:
  masquerade: no
  forward-ports:
  source-ports:
  icmp-blocks:
  rich rules:

[root@rhel1 ~]#
```

Figure 9.11 – Firewall commands utilized to allow service connectivity

The Ansible playbook for setting up `firewalld` services is as follows:

```
- name: firewalld configuration
  hosts: all
  become: true
  become_method: sudo

  tasks:
    - name: Enable Firewalld services
      firewalld:
        permanent: yes
        immediate: yes
        service: "{{ item }}"
        state: enabled
      loop:
```

```
            - http
            - https


    - name: Reload firewall
      command:
        cmd: firewall-cmd --reload


    - name: Show open services and ports
      command:
        cmd: firewall-cmd --list-all
```

Please see the following screenshot with the inventory for your reference:

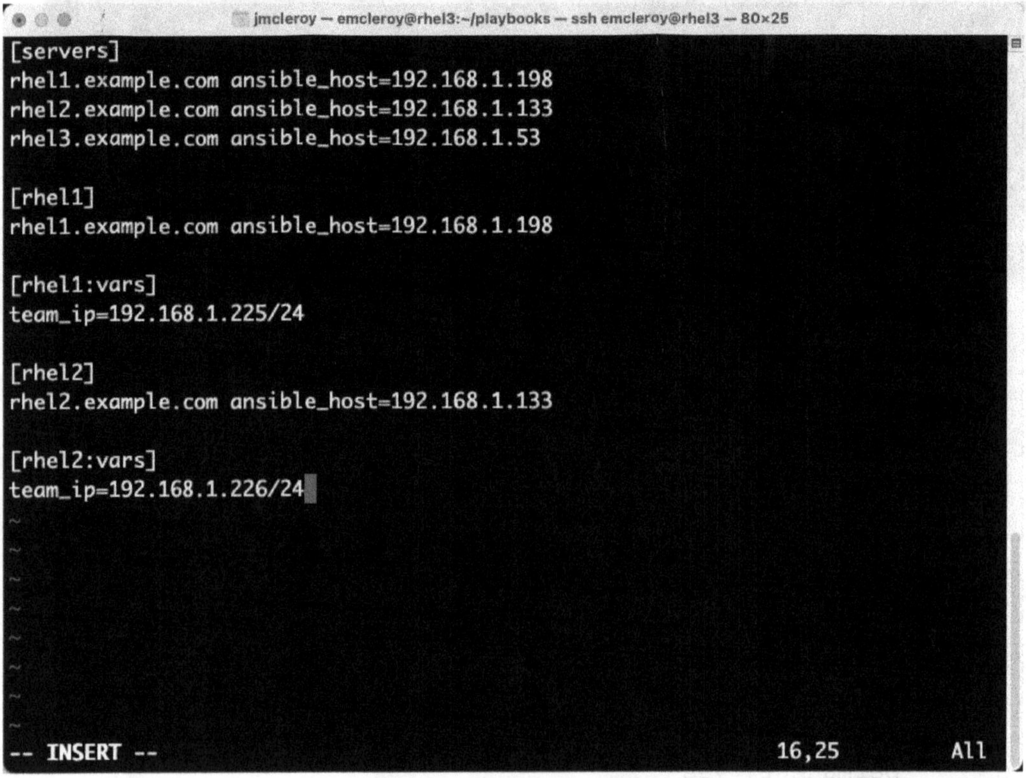

```
jmcleroy — emcleroy@rhel3:~/playbooks — ssh emcleroy@rhel3 — 80×25
[servers]
rhel1.example.com ansible_host=192.168.1.198
rhel2.example.com ansible_host=192.168.1.133
rhel3.example.com ansible_host=192.168.1.53

[rhel1]
rhel1.example.com ansible_host=192.168.1.198

[rhel1:vars]
team_ip=192.168.1.225/24

[rhel2]
rhel2.example.com ansible_host=192.168.1.133

[rhel2:vars]
team_ip=192.168.1.226/24
~
~
~
~
~
~
-- INSERT --                                          16,25            All
```

Figure 9.12 – Playbook inventory for all labs

After we have set up and run the playbook with `ansible-playbook -i inventory firewalld.yml -u emcleroy -k –ask-become -v`, we will see the output in the following screenshot:

Figure 9.13 – Successful Ansible playbook completion

Question 2 – For this question, we would like you to provide access to HTTP via the 192.168.1.0/24 domain, but block access via the 172.16.1.0/24 domain.

Please do this manually and also set it up in an Ansible Automation playbook.

Answer 2 – We will use firewall-cmd rich rules to accomplish this change. We will use the following commands to successfully complete this exercise:

```
[emcleroy@rhel1 ~]$ sudo firewall-cmd --permanent --add-rich-
rule='rule family=ipv4 source address=192.168.1.0/24 service
name=http accept'
success
[emcleroy@rhel1 ~]$ sudo firewall-cmd --permanent --add-rich-
rule='rule family=ipv4 source address=172.16.1.0/24 service
name=http reject'
success
```

```
[emcleroy@rhel1 ~]$ sudo firewall-cmd --list-rich-rules
rule family="ipv4" source address="192.168.1.0/24" service
name="http" accept
rule family="ipv4" source address="172.16.1.0/24" service
name="http" reject
[emcleroy@rhel1 ~]$ sudo firewall-cmd --reload
success
```

Now, let's create an Ansible playbook to accomplish the same changes to the firewall rules, which can be seen in the following playbook:

```
- name: Firewalld Rich Rule
  hosts: all
  become: true
  become_method: sudo

  tasks:
    - name: Set rich rules up on firewall
      firewalld:
        permanent: yes
        immediate: yes
        rich_rule: "{{ item }}"
        state: enabled
      loop:
        - 'rule family=ipv4 source address=192.168.1.0/24
service name="http"  accept'
        - 'rule family=ipv4 source address=172.16.1.0/24 service
name="http" reject'

    - name: Show rich rules
      command:
        cmd: firewall-cmd --list-rich-rules

    - name: Reload firewall
      command:
        cmd: firewall-cmd --reload
```

After we have run the playbook using `ansible-playbook -i inventory richrule.yml -u emcleroy -k –ask-become -v`, we will see the output in the following screenshot:

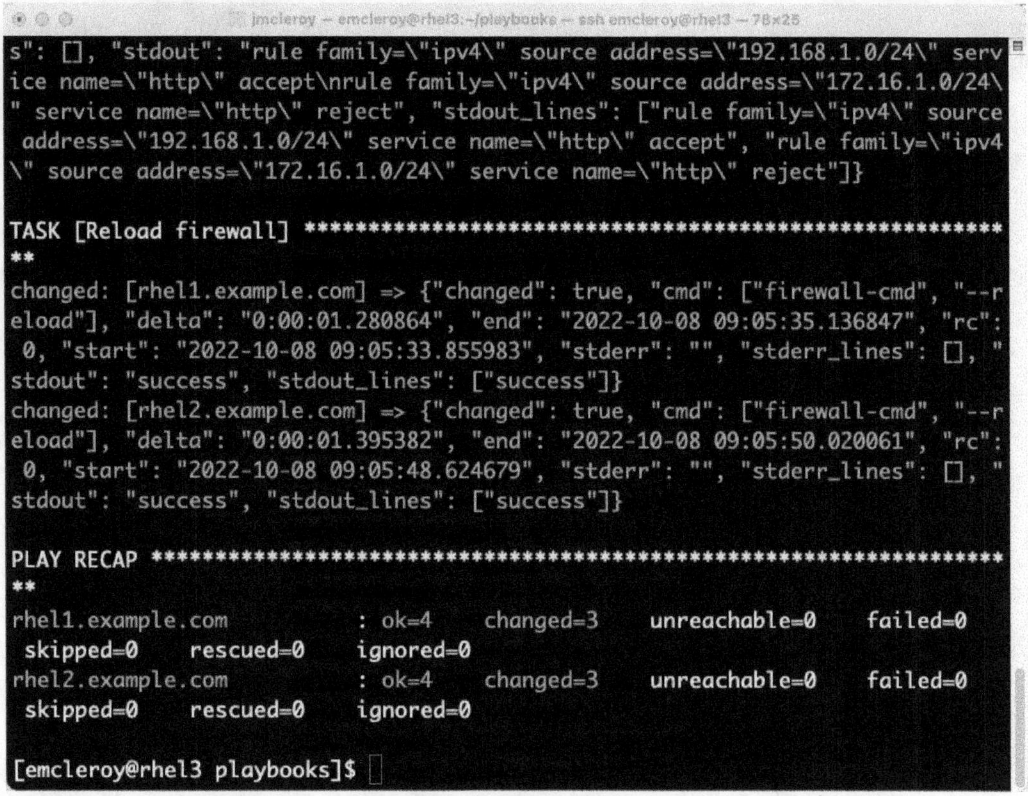
jmcleroy — emcleroy@rhel3:~/playbooks — ssh emcleroy@rhel3 — 78x25
s": [], "stdout": "rule family=\"ipv4\" source address=\"192.168.1.0/24\" serv
ice name=\"http\" accept\nrule family=\"ipv4\" source address=\"172.16.1.0/24\
" service name=\"http\" reject", "stdout_lines": ["rule family=\"ipv4\" source
address=\"192.168.1.0/24\" service name=\"http\" accept", "rule family=\"ipv4
\" source address=\"172.16.1.0/24\" service name=\"http\" reject"]}
TASK [Reload firewall] ***
**
changed: [rhel1.example.com] => {"changed": true, "cmd": ["firewall-cmd", "--r
eload"], "delta": "0:00:01.280864", "end": "2022-10-08 09:05:35.136847", "rc":
0, "start": "2022-10-08 09:05:33.855983", "stderr": "", "stderr_lines": [], "
stdout": "success", "stdout_lines": ["success"]}
changed: [rhel2.example.com] => {"changed": true, "cmd": ["firewall-cmd", "--r
eload"], "delta": "0:00:01.395382", "end": "2022-10-08 09:05:50.020061", "rc":
0, "start": "2022-10-08 09:05:48.624679", "stderr": "", "stderr_lines": [], "
stdout": "success", "stdout_lines": ["success"]}
PLAY RECAP ***
**
rhel1.example.com : ok=4 changed=3 unreachable=0 failed=0
skipped=0 rescued=0 ignored=0
rhel2.example.com : ok=4 changed=3 unreachable=0 failed=0
skipped=0 rescued=0 ignored=0
[emcleroy@rhel3 playbooks]$

Figure 9.14 – Successful Ansible playbook completion output

Managing SELinux

In this section, we will be working to fix SELinux issues when creating files and folders to ensure there are no permission issues.

Question 1 – For this question, we would like you to create a folder and text file in the root directory and then repair the SELinux contexts to the correct permissions:

```
Folder: /srv/test
File: /srv/test/text.txt
File body: Testing SELinux contexts.
```

Please do this manually and also set it up in an Ansible Automation playbook.

Answer 1 – We will use the following commands to successfully create and resolve any SELinux context issues:

```
[emcleroy@rhel1 ~]$ sudo mkdir /srv/test
[emcleroy@rhel1 ~]$ sudo vi /srv/test/text.txt
[emcleroy@rhel1 ~]$ restorecon -Rv /srv/test
[emcleroy@rhel1 ~]$ ls -lZ /srv/test
total 4
-rw-r--r--. 1 root root unconfined_u:object_r:var_t:s0 33
Oct  8 09:16 text.txt
```

Next, we will accomplish this action with an Ansible playbook as follows:

```
- name: SELinux training
  hosts: all
  become: true
  become_method: sudo

  tasks:
    - name: Create directory
      file:
        path: /srv/test
        state: directory
        mode: '0755'

    - name: Create a file in the directory
      copy:
        dest: "/srv/test/text.txt"
        content: |
          Thank you for reading my book!!!

    - name: Restore any SELinux issues if they exist
      command:
        cmd: restorecon -Rv /srv/test

    - name: View attributes
      command:
        cmd: ls -lZ /srv/test
```

After we have created the Ansible playbook, we will run it with this command: `ansible-playbook -i inventory selinux.yml -u emcleroy -k --ask-become -v`. You can see the results in the following screenshot:

```
[], "stdout": "", "stdout_lines": []}

TASK [View attributes] ***************************************************
**
changed: [rhel1.example.com] => {"changed": true, "cmd": ["ls", "-lZ", "/srv/t
est"], "delta": "0:00:00.003553", "end": "2022-10-08 09:33:10.755269", "rc": 0
, "start": "2022-10-08 09:33:10.751716", "stderr": "", "stderr_lines": [], "st
dout": "total 4\n-rw-r--r--. 1 root root unconfined_u:object_r:var_t:s0 33 Oct
  8 09:16 text.txt", "stdout_lines": ["total 4", "-rw-r--r--. 1 root root unco
nfined_u:object_r:var_t:s0 33 Oct  8 09:16 text.txt"]}
changed: [rhel2.example.com] => {"changed": true, "cmd": ["ls", "-lZ", "/srv/t
est"], "delta": "0:00:00.003400", "end": "2022-10-08 09:33:25.526241", "rc": 0
, "start": "2022-10-08 09:33:25.522841", "stderr": "", "stderr_lines": [], "st
dout": "total 4\n-rw-r--r--. 1 root root system_u:object_r:var_t:s0 33 Oct  8
09:33 text.txt", "stdout_lines": ["total 4", "-rw-r--r--. 1 root root system_u
:object_r:var_t:s0 33 Oct  8 09:33 text.txt"]}

PLAY RECAP ***************************************************************
**
rhel1.example.com          : ok=5    changed=2    unreachable=0    failed=0
 skipped=0    rescued=0    ignored=0
rhel2.example.com          : ok=5    changed=3    unreachable=0    failed=0
 skipped=0    rescued=0    ignored=0

[emcleroy@rhel3 playbooks]$
```

Figure 9.15 – Successful Ansible Automation playbook run output

We learned how to resolve SELinux issues with files and folders in this section with easy-to-remember commands that should address issues you run into in the exam. Next, we will showcase how to control system processes that are running on your servers.

Managing system processes

In this section, we will start and enable services. We will also install packages, as they are needed for us to start and enable a service that is not already running. We are using `postgresql` as an example, but this will work for all the services you install, such as DHCP, DNS, and SMB.

Question 1 – For this question, we would like you to install `postgresql` and start and enable it on `rhel1`.

Please do this manually and also set it up in an Ansible Automation playbook.

Answer 1 – We will use the following commands to install, enable, and start `postgresql` on `rhel1`:

```
[emcleroy@rhel1 ~]$ sudo dnf install @postgresql
[emcleroy@rhel1 ~]$ sudo systemctl enable postgresql
[emcleroy@rhel1 ~]$ sudo postgresql-setup --initdb
 * Initializing database in '/var/lib/pgsql/data'
 * Initialized, logs are in /var/lib/pgsql/initdb_postgresql.
log
[emcleroy@rhel1 ~]$ sudo systemctl start postgresql
```

After we have completed the installation and startup of `postgresql` manually, we need to uninstall `postgresql` so that Ansible can install it. We will use the following command to remove `postgresql` from `rhel1`:

```
sudo dnf remove @postgresql
```

Once `postgresql` has been removed, we will set up an Ansible playbook as follows to install, enable, and start the `postgresql` service:

```
- name: Install package, enable and start it.
  hosts: rhel1.example.com
  become: true
  become_method: sudo

  tasks:
    - name: Install postgresql
      command:
        cmd: "dnf install @postgresql -y"

    - name: Initialize the database
      command:
        cmd: postgresql-setup --initdb
      ignore_errors: true

    - name: Enable and start postgresql service
      service:
        name: postgresql
        state: started
```

```
        enabled: true

    - name: Check to ensure service is started
      command:
        cmd: systemctl status postgresql
```

We have added `ignore_errors` to the initialization step due to possible failures that do not affect the starting or enabling of the service.

After we have completed writing the playbook, we will run it using the `ansible-playbook -i inventory postgresql.yml -u emcleroy -k --ask-become -v` command. The result of running the playbook is shown in the following screenshot:

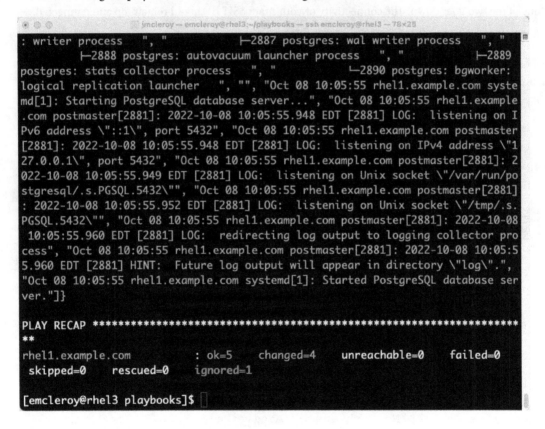

Figure 9.16 – Successful run of the postgresql playbook

In this section, we learned how to install, start, and enable services with the `postgresql` service as an example. These commands can be used with all services that you can install on an RHEL 8.1 system. In the next section, we will delve into network teaming.

Managing link aggregation

In this section, we will set up network teaming:

Question 1 – For this question, we would like you to set up network teaming on the additional NICs that were provisioned in the setup process:

```
Rhel1 IPv4: 192.168.1.225
Rhel2 IPv4: 192.168.1.226
```

My setup has the NICs as **enp0s8** and **enp0s9** – yours might be different depending on your setup. Set up the connectivity as round-robin runner. You should be able to connect to each server using these IPs once set up. You should have connectivity to the network team when both NICs are up and when one of the two goes down. The connectivity of your main SSH IP should remain up and connected.

Please do this manually and also set it up in an Ansible Automation playbook.

Answer 1 – We will be using the following commands to set up the team on both systems. Please make sure you use the correct IPv4 address per server:

```
[emcleroy@rhel1 ~]$ sudo nmcli con add type team con-name team1
ifname team1 team.runner roundrobin
Connection 'team1' (58770e3c-89bf-4a87-9b26-e9ad581ef978)
successfully added.
[emcleroy@rhel1 ~]$ sudo nmcli con add type ethernet slave-type
team con-name team1-enp0s8 ifname enp0s8 master team1
Connection 'team1-enp0s8' (3886d9f8-2db6-4269-9d9b-
2973b24f74ce) successfully added.
[emcleroy@rhel1 ~]$ sudo nmcli con add type ethernet slave-type
team con-name team1-enp0s9 ifname enp0s9 master team1
Connection 'team1-enp0s9' (d845c4e1-b354-4c58-904d-
f9cf0c3d1e59) successfully added.
[emcleroy@rhel1 ~]$ sudo nmcli con mod team1 ipv4.addresses
192.168.1.225/24
[emcleroy@rhel1 ~]$ sudo nmcli con mod team1 ipv4.method manual
[emcleroy@rhel1 ~]$ sudo ifdown team1
Connection 'team1' successfully deactivated (D-Bus active path:
/org/freedesktop/NetworkManager/ActiveConnection/2)
[emcleroy@rhel1 ~]$ sudo ifup team1
Connection successfully activated (master waiting for
slaves) (D-Bus active path: /org/freedesktop/NetworkManager/
ActiveConnection/5)
[emcleroy@rhel1 ~]$ ifconfig
```

```
team1: flags=4163<UP,BROADCAST,RUNNING,MULTICAST>  mtu 1500
        inet 192.168.1.225  netmask 255.255.255.0  broadcast
192.168.1.255
        inet6 fe80::4036:4e32:d1af:fd0a  prefixlen 64  scopeid
0x20<link>
        ether 08:00:27:ad:b3:d8  txqueuelen 1000  (Ethernet)
        RX packets 159  bytes 8184 (7.9 KiB)
        RX errors 0  dropped 30  overruns 0  frame 0
        TX packets 8  bytes 688 (688.0 B)
        TX errors 0  dropped 0 overruns 0  carrier
0  collisions 0
```

Now, repeat the process for Rhel2. Once that is complete, you should have a team1 interface that is up and running with roundrobin connectivity. You can use ping to test the interface, as shown in the following screenshot:

```
[emcleroy@rhel2 ~]$ ping 192.168.1.225
PING 192.168.1.225 (192.168.1.225) 56(84) bytes of data.
64 bytes from 192.168.1.225: icmp_seq=1 ttl=64 time=0.998 ms
64 bytes from 192.168.1.225: icmp_seq=2 ttl=64 time=0.452 ms
64 bytes from 192.168.1.225: icmp_seq=3 ttl=64 time=0.499 ms
^C
--- 192.168.1.225 ping statistics ---
3 packets transmitted, 3 received, 0% packet loss, time 15ms
rtt min/avg/max/mdev = 0.452/0.649/0.998/0.248 ms
[emcleroy@rhel2 ~]$
```

Figure 9.17 – ping tests to ensure connectivity to the newly set up team

After we have completed the team creation manually, we will set up an Ansible playbook as follows to create the teams:

```
- name: Create a network team
  hosts: rhel1.example.com, rhel2.example.com
  become: true
  become_method: sudo

  tasks:
    - name: Create network team
      command:
        cmd: nmcli con add type team con-name team1 ifname
team1 team.runner roundrobin

    - name: Add slave interfaces
      command:
        cmd: 'nmcli con add type ethernet slave-type team con-
name team1-"{{ item }}" ifname "{{ item }}" master team1'
      loop:
        - enp0s8
        - enp0s9

    - name: Add Static IP to team1
      command:
        cmd: nmcli con mod team1 ipv4.addresses "{{ team_ip }}"

    - name: Add interface method
      command:
        cmd: nmcli con mod team1 ipv4.method manual
```

We will then add host variables to our inventory so that each server gets the correct IP, as shown in the following screenshot:

```
[servers]
rhel1.example.com ansible_host=192.168.1.198
rhel2.example.com ansible_host=192.168.1.133

[rhel1]
rhel1.example.com ansible_host=192.168.1.198

[rhel1:vars]
team_ip=192.168.1.225/24

[rhel2]
rhel2.example.com ansible_host=192.168.1.133

[rhel2:vars]
team_ip=192.168.1.226/24
~
~
~
~
~
~
~
"inventory" 15L, 286C                                    1,1            All
```

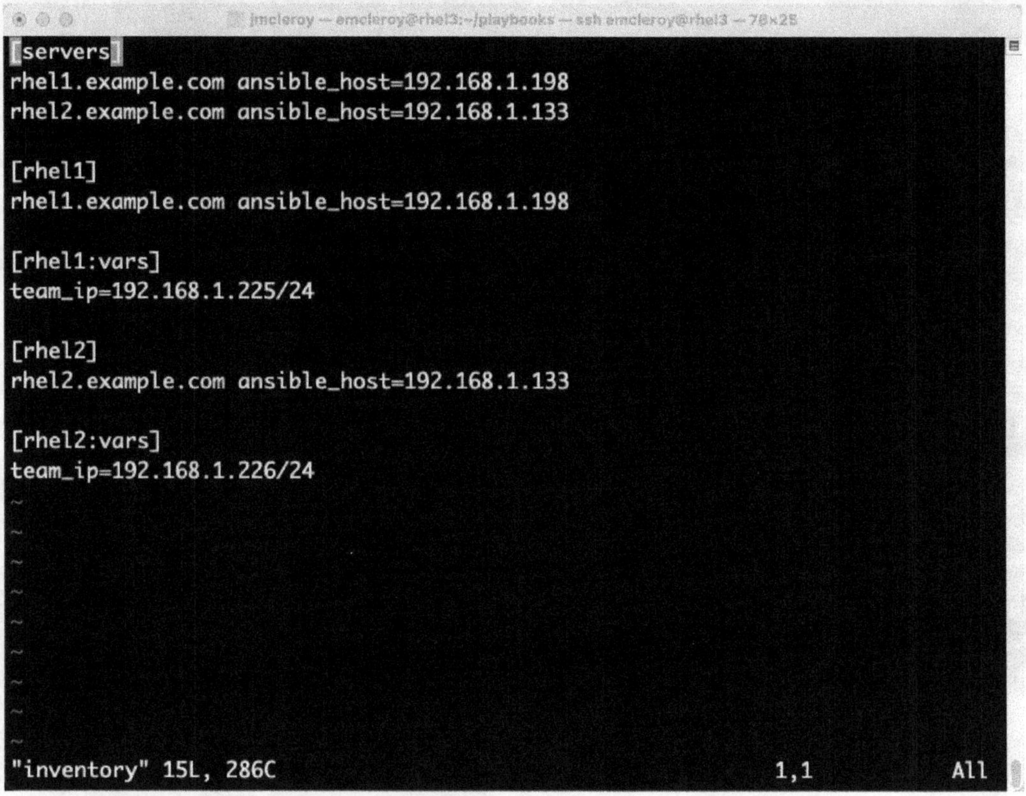

Figure 9.18 – Updated inventory file with host variables

After we have created the playbook, we can run it using the `ansible-playbook -i inventory networkteam.yml -u emcleroy -k --ask-become -v` command – the partial output is shown in the following screenshot:

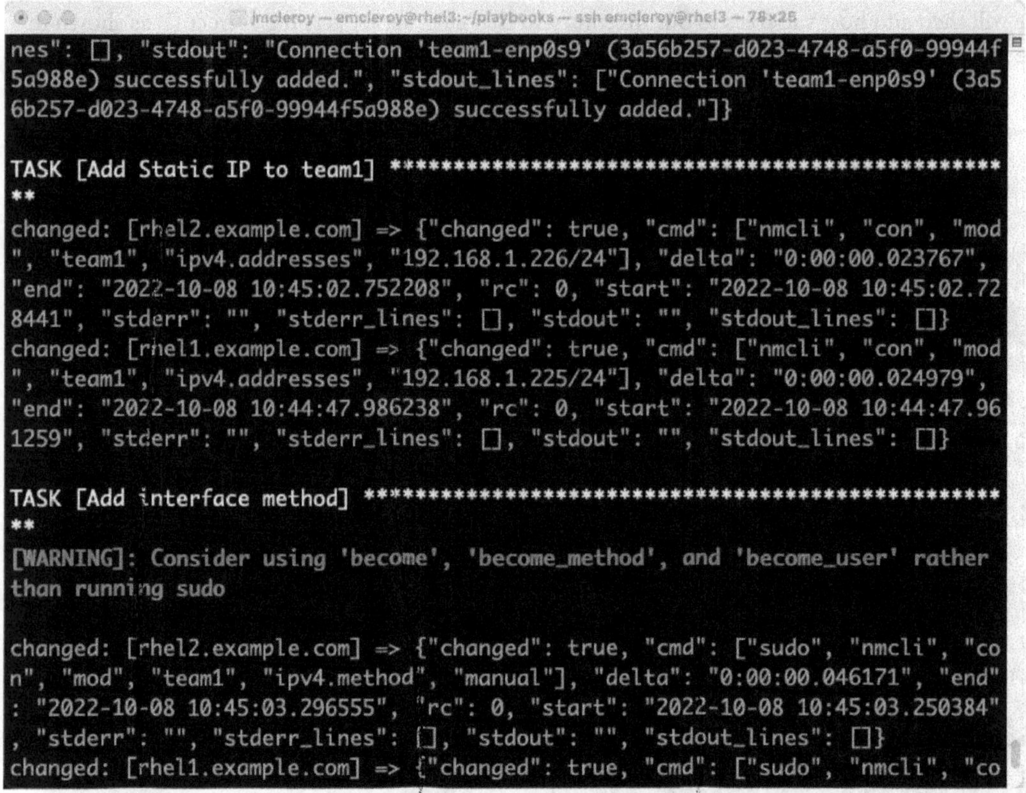

nes": [], "stdout": "Connection 'team1-enp0s9' (3a56b257-d023-4748-a5f0-99944f
5a988e) successfully added.", "stdout_lines": ["Connection 'team1-enp0s9' (3a5
6b257-d023-4748-a5f0-99944f5a988e) successfully added."]}

TASK [Add Static IP to team1] ***
**
changed: [rhel2.example.com] => {"changed": true, "cmd": ["nmcli", "con", "mod
", "team1", "ipv4.addresses", "192.168.1.226/24"], "delta": "0:00:00.023767",
"end": "2022-10-08 10:45:02.752208", "rc": 0, "start": "2022-10-08 10:45:02.72
8441", "stderr": "", "stderr_lines": [], "stdout": "", "stdout_lines": []}
changed: [rhel1.example.com] => {"changed": true, "cmd": ["nmcli", "con", "mod
", "team1", "ipv4.addresses", "192.168.1.225/24"], "delta": "0:00:00.024979",
"end": "2022-10-08 10:44:47.986238", "rc": 0, "start": "2022-10-08 10:44:47.96
1259", "stderr": "", "stderr_lines": [], "stdout": "", "stdout_lines": []}

TASK [Add interface method] ***
**
[WARNING]: Consider using 'become', 'become_method', and 'become_user' rather
than running sudo

changed: [rhel2.example.com] => {"changed": true, "cmd": ["sudo", "nmcli", "co
n", "mod", "team1", "ipv4.method", "manual"], "delta": "0:00:00.046171", "end"
: "2022-10-08 10:45:03.296555", "rc": 0, "start": "2022-10-08 10:45:03.250384"
, "stderr": "", "stderr_lines": [], "stdout": "", "stdout_lines": []}
changed: [rhel1.example.com] => {"changed": true, "cmd": ["sudo", "nmcli", "co

Figure 9.19 – Successful network teaming playbook run

In this section, we learned how to team network interfaces together to provide redundancy across multiple NICs. In the next section, we will set up and manage DNS services.

Managing DNS

In this section, we will set up a DNS server.

Question 1 – For this question, we would like you to set up a DNS server. We would like you to add forward and reverse records for IPv4. We will set up the DNS server on `rhel1.example.com`.

A and reverse aka PTR records will use the following information:

```
rhel1.example.com - 192.168.1.198
rhel2.example.com - 192.168.1.133
rhel3.example.com - 192.168.1.53
```

Answer 1 – We will start by installing the `bind` package:

```
[emcleroy@rhel1 ~]$ sudo dnf install bind -y
```

Next, we will set up the firewall rules to allow for connectivity and the delivery of DNS records:

```
[emcleroy@rhel1 ~]$ sudo firewall-cmd --permanent --add-
service=dns
success
[emcleroy@rhel1 ~]$ sudo firewall-cmd --reload
Success
```

From installing the package to setting up firewall rules, we will now configure the config DNS file, `/etc/named.conf`. This file should look like the following screenshot:

```
options {
        listen-on port 53 { 127.0.0.1; 192.168.1.198; };
        listen-on-v6 port 53 { ::1; };
        directory       "/var/named";
        dump-file       "/var/named/data/cache_dump.db";
        statistics-file "/var/named/data/named_stats.txt";
        memstatistics-file "/var/named/data/named_mem_stats.txt";
        secroots-file   "/var/named/data/named.secroots";
        recursing-file  "/var/named/data/named.recursing";
        allow-query     { localhost; 192.168.1.198; };

        recursion no;

        dnssec-enable yes;
        dnssec-validation yes;

        managed-keys-directory "/var/named/dynamic";

        pid-file "/run/named/named.pid";
        session-keyfile "/run/named/session.key";

        /* https://fedoraproject.org/wiki/Changes/CryptoPolicy */
        include "/etc/crypto-policies/back-ends/bind.config";
};
```

Figure 9.20 – Configuration example settings

Make sure to include the new zones in the configuration file as shown in the following screenshot:

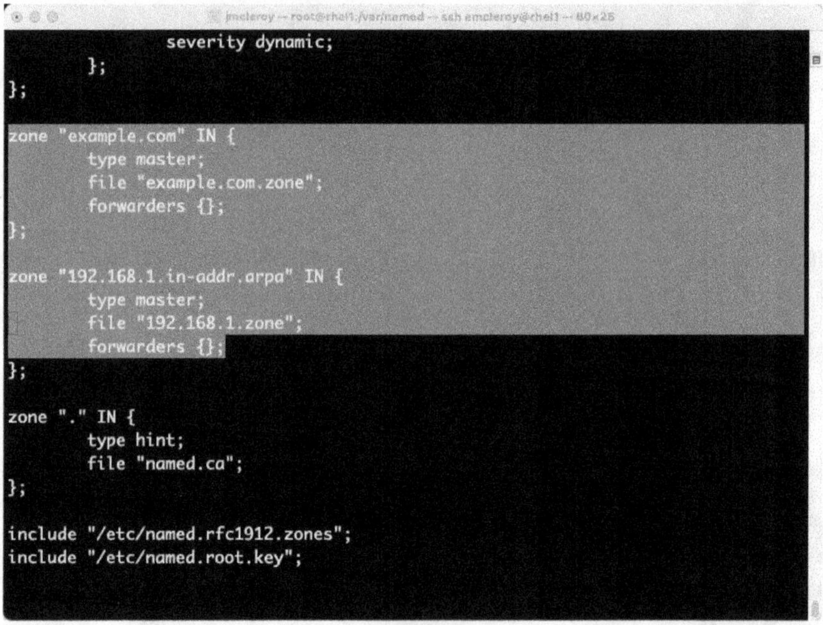

```
                severity dynamic;
        };
};

zone "example.com" IN {
        type master;
        file "example.com.zone";
        forwarders {};
};

zone "192.168.1.in-addr.arpa" IN {
        type master;
        file "192.168.1.zone";
        forwarders {};
};

zone "." IN {
        type hint;
        file "named.ca";
};

include "/etc/named.rfc1912.zones";
include "/etc/named.root.key";
```

Figure 9.21 – Zone information in the configuration file

Next, we need to create the zone files in the /var/named directory. These will include the forward and reverse zone records of 192.168.1.zone and example.com.zone. We can see an example of each one in the following screenshots:

```
$TTL 3H
@       IN SOA rhel1.example.com. admin.example.com (
2; serial
65; refresh
75; retry
8000; expire
60) ; minimum

        NS      rhel1
        A       127.0.0.1
        AAAA    ::1

rhel1       IN  A 192.168.1.198
rhel2       IN  A 192.168.1.133
rhel3       IN  A 192.168.1.53
~
~
~
~
~
~
~
~
~
"/var/named/example.com.zone" 15L, 289C
```

Figure 9.22 – DNS zone file setup

After we have set up the forward zone, we will set up the reverse zone, as shown in the following screenshot:

```
$TTL 3H
@       IN SOA rhel1.example.com. admin.example.com. (
                                        2       ; serial
                                        8000      ; refresh
                                        9000      ; retry
                                        10000      ; expire
                                        5000)     ; minimum
        NS       rhel1.example.com.

198        IN  PTR  rhel1.example.com.
133        IN  PTR  rhel2.example.com.
53         IN  PTR  rhel3.example.com.
~
~
~
~
~
~
~
~
~
~
~
~
~
"/var/named/192.168.1.zone" 12L, 510C
```

Figure 9.23 – Reverse zone file

Next, we will start and enable the `named` service that the `bind` package installed for running the DNS server:

```
[root@rhel1 named]# systemctl enable named
Created symlink /etc/systemd/system/multi-user.target.wants/
named.service → /usr/lib/systemd/system/named.service.
[root@rhel1 named]# systemctl start named
```

Then, we will do a lookup to ensure that we are getting the right addresses back for the names:

```
[root@rhel1 named]# nslookup rhel2.example.com 192.168.1.198
Server:         192.168.1.198
Address:    192.168.1.198#53


Name: rhel2.example.com
Address: 192.168.1.133
```

Next, we will install this by using Ansible Automation with the following playbook and templates:

```
---
- name: Install and configure DNS
  hosts: rhel1.example.com
  become: true
  become_method: sudo

  tasks:
    - name: Install DNS server
      package:
        name: bind
        state: latest

    - name: Move bind configuration to named.conf
      template:
        src: "{{ playbook_dir }}/named.conf.j2"
        dest: "/etc/named.conf"

    - name: Move example.com.zone and 192.168.1.zone to /var/
named
      template:
```

```
      src: "{{ playbook_dir}}/{{ item }}.j2"
      dest: "/var/named/{{ item }}"
    loop:
      - 'example.com.zone'
      - '192.168.1.zone'

  - name: Set firewall rules
    firewalld:
      service: dns
      permanent: true
      state: enabled

  - name: Reload firewall
    command:
      cmd: firewall-cmd --reload

  - name: Start and enable the DNS service
    service:
      name: named
      state: restarted
      enabled: true
```

The following are the templates that need to be in place for the correct configuration files to be successfully modified during the playbook execution as well. First, we have the named.conf.j2 configuration file:

```
//
// named.conf
//
// Provided by Red Hat bind package to configure the ISC BIND
named(8) DNS
// server as a caching only nameserver (as a localhost DNS
resolver only).
//
// See /usr/share/doc/bind*/sample/ for example named
configuration files.
//
```

```
options {
      listen-on port 53 { 127.0.0.1; 192.168.1.198; };
      listen-on-v6 port 53 { ::1; };
      directory   "/var/named";
      dump-file   "/var/named/data/cache_dump.db";
      statistics-file "/var/named/data/named_stats.txt";
      memstatistics-file "/var/named/data/named_mem_stats.txt";
      secroots-file    "/var/named/data/named.secroots";
      recursing-file   "/var/named/data/named.recursing";
      allow-query      { localhost; 192.168.1.198; };

      recursion no;

      dnssec-enable yes;
      dnssec-validation yes;

      managed-keys-directory "/var/named/dynamic";

      pid-file "/run/named/named.pid";
      session-keyfile "/run/named/session.key";

      /* https://fedoraproject.org/wiki/Changes/CryptoPolicy */
      include "/etc/crypto-policies/back-ends/bind.config";
};

logging {
        channel default_debug {
                file "data/named.run";
                severity dynamic;
        };
};

zone "example.com" IN {
        type master;
        file "example.com.zone";
        forwarders {};
```

```
};

zone "192.168.1.in-addr.arpa" IN {
        type master;
        file "192.168.1.zone";
        forwarders {};
};

zone "." IN {
     type hint;
     file "named.ca";
};

include "/etc/named.rfc1912.zones";
include "/etc/named.root.key";
```

Then, we have the zone files that are needed in order to provide the DNS records. First up is the example.com.zone.j2 file:

```
$TTL 3H
@        IN SOA rhel1.example.com. admin.example.com (
2; serial
65; refresh
75; retry
8000; expire
60) ; minimum


         NS      rhel1
         A       127.0.0.1
         AAAA    ::1

rhel1         IN  A 192.168.1.198
rhel2         IN  A 192.168.1.133
rhel3         IN  A 192.168.1.53
```

Finally, we will add in `192.168.1.zone.j2` for the **Pointer Record (PTR)** records, also known as the reverse lookup zone file:

```
$TTL 3H
@         IN SOA rhel1.example.com. admin.example.com. (
                                        2        ; serial
                                        8000       ; refresh
                                        9000       ; retry
                                        10000       ; expire
                                        5000)    ; minimum
          NS       rhel1.example.com.


198        IN   PTR   rhel1.example.com.
133        IN   PTR   rhel2.example.com.
53         IN   PTR   rhel3.example.com.
```

After we create the playbook, we run the `ansible-playbook -i /home/emcleroy/playbooks/inventory dns.yml -u emcleroy -k --ask-become -v` command, and we can see the successful output in the following screenshot:

Figure 9.24 – Successful DNS server deployment via Ansible Automation

Managing DHCP

In this section, we will be setting up a DHCP server.

Question 1 – For this question, we would like you to set up a DHCP server on `rhel1.example.com` that serves the `192.168.1.0/24` subnet with an available IP range of `192.168.1.100 - 192.168.1.220`. The DNS server is `192.168.1.198` or your `rhel1.example.com` IP address. We will want static entries for the following:

```
rhel4.example.com MAC Address: 08:00:27:AB:81:10 IP:
192.168.1.120
rhel5.example.com MAC Address: 08:00:27:AB:81:11 IP:
192.168.1.121
```

Answer 1 – We will start by installing the `dhcp-server` package as shown in the following:

```
[emcleroy@rhel1 ~]$ sudo dnf install dhcp-server -y
```

Next, we will set up the dhcp file to accommodate and provide the IP range in `/etc/dhcp/dhcpd.conf`, as seen in the following screenshot:

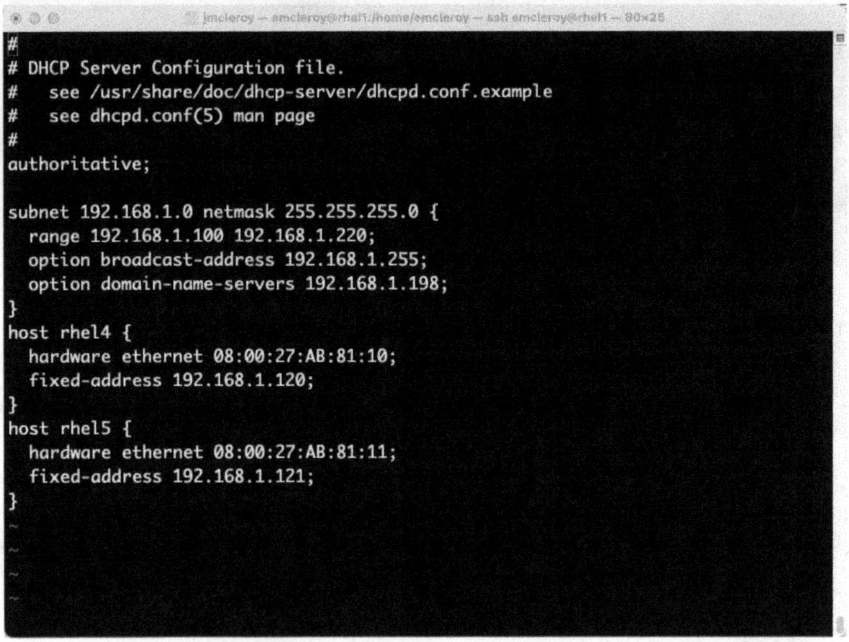

```
#
# DHCP Server Configuration file.
#   see /usr/share/doc/dhcp-server/dhcpd.conf.example
#   see dhcpd.conf(5) man page
#
authoritative;

subnet 192.168.1.0 netmask 255.255.255.0 {
  range 192.168.1.100 192.168.1.220;
  option broadcast-address 192.168.1.255;
  option domain-name-servers 192.168.1.198;
}
host rhel4 {
  hardware ethernet 08:00:27:AB:81:10;
  fixed-address 192.168.1.120;
}
host rhel5 {
  hardware ethernet 08:00:27:AB:81:11;
  fixed-address 192.168.1.121;
}
```

Figure 9.25 – DHCP server configuration file

Remember there is no need to remember all the lines, as there is an example provided after the installation found here: `/usr/share/doc/dhcp-server/dhcpd.conf.example`. After we have set up the DHCP configuration file, we will open the firewall rules and start and enable dhcp services for the server:

```
[emcleroy@rhel1 ~]$ sudo firewall-cmd --permanent --add-
service=dhcp
[emcleroy@rhel1 ~]$ firewall-cmd -reload
[emcleroy@rhel1 ~]$ sudo systemctl start dhcpd
[emcleroy@rhel1 ~]$ sudo systemctl enable dhcpd
Created symlink /etc/systemd/system/multi-user.target.wants/
dhcpd.service → /usr/lib/systemd/system/dhcpd.service.
```

Now, let's set up a DHCP server using Ansible Automation. We will start with the following playbook:

```
---
- name: Install and configure DHCP
  hosts: rhel1.example.com
  become: true
  become_method: sudo

  tasks:
    - name: Install dhcp server
      package:
        name: dhcp-server
        state: latest

    - name: Copy the dhcpd.conf file to the server
      template:
        src: "{{ playbook_dir }}/dhcpd.conf.j2"
        dest: /etc/dhcp/dhcpd.conf

    - name: Open firewall rules
      firewalld:
        service: dhcp
        permanent: true
        state: enabled

    - name: Reload firewall rules
      command:
```

```
        cmd: firewall-cmd --reload

    - name: Start and enable dhcp server
      service:
        name: dhcpd
        enabled: true
        state: restarted
```

The following code displays the contents of the dhcpd.conf.j2 file:

```
#
# DHCP Server Configuration file.
#   see /usr/share/doc/dhcp-server/dhcpd.conf.example
#   see dhcpd.conf(5) man page
#
authoritative;

subnet 192.168.1.0 netmask 255.255.255.0 {
  range 192.168.1.100 192.168.1.220;
  option broadcast-address 192.168.1.255;
  option domain-name-servers 192.168.1.198;
}
host rhel4 {
  hardware ethernet 08:00:27:AB:81:10;
  fixed-address 192.168.1.120;
}
host rhel5 {
  hardware ethernet 08:00:27:AB:81:11;
  fixed-address 192.168.1.121;
}
```

After we have created the playbook, we can run it using the ansible-playbook -i inventory dhcp.yml -u emcleroy -k --ask-become -v command to run an Ansible playbook successfully, as can be seen in the following screenshot:

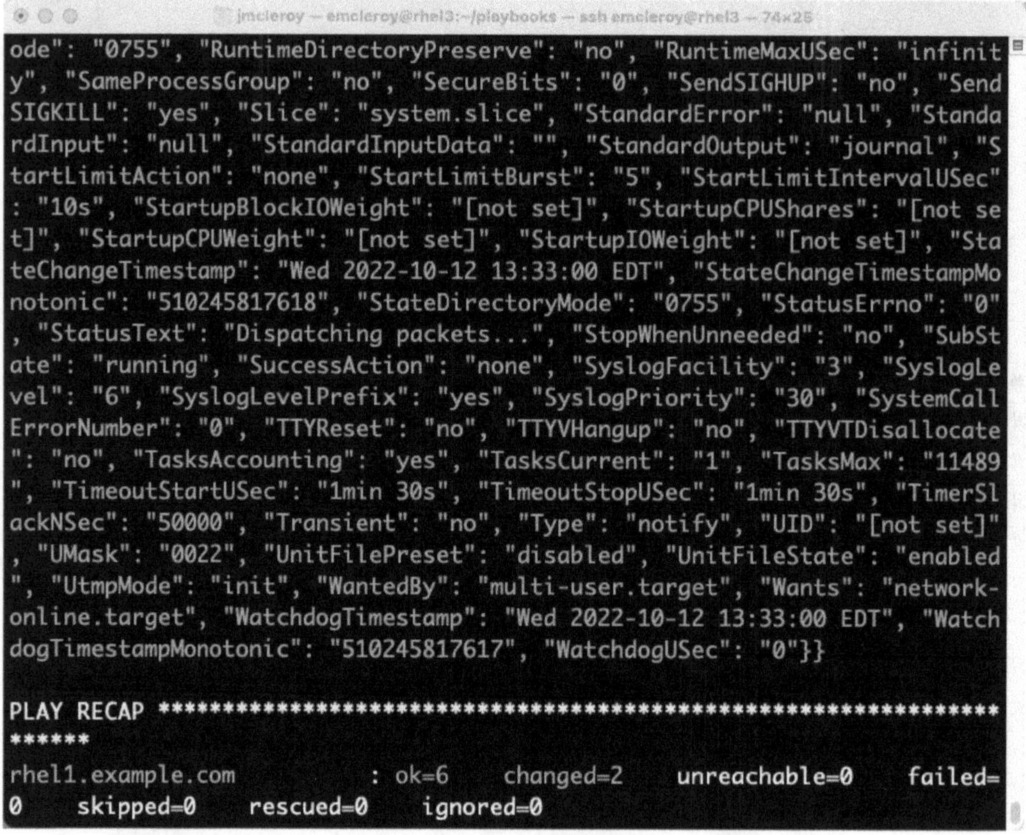

ode": "0755", "RuntimeDirectoryPreserve": "no", "RuntimeMaxUSec": "infinit
y", "SameProcessGroup": "no", "SecureBits": "0", "SendSIGHUP": "no", "Send
SIGKILL": "yes", "Slice": "system.slice", "StandardError": "null", "Standa
rdInput": "null", "StandardInputData": "", "StandardOutput": "journal", "S
tartLimitAction": "none", "StartLimitBurst": "5", "StartLimitIntervalUSec"
: "10s", "StartupBlockIOWeight": "[not set]", "StartupCPUShares": "[not se
t]", "StartupCPUWeight": "[not set]", "StartupIOWeight": "[not set]", "Sta
teChangeTimestamp": "Wed 2022-10-12 13:33:00 EDT", "StateChangeTimestampMo
notonic": "510245817618", "StateDirectoryMode": "0755", "StatusErrno": "0"
, "StatusText": "Dispatching packets...", "StopWhenUnneeded": "no", "SubSt
ate": "running", "SuccessAction": "none", "SyslogFacility": "3", "SyslogLe
vel": "6", "SyslogLevelPrefix": "yes", "SyslogPriority": "30", "SystemCall
ErrorNumber": "0", "TTYReset": "no", "TTYVHangup": "no", "TTYVTDisallocate
": "no", "TasksAccounting": "yes", "TasksCurrent": "1", "TasksMax": "11489
", "TimeoutStartUSec": "1min 30s", "TimeoutStopUSec": "1min 30s", "TimerSl
ackNSec": "50000", "Transient": "no", "Type": "notify", "UID": "[not set]"
, "UMask": "0022", "UnitFilePreset": "disabled", "UnitFileState": "enabled
", "UtmpMode": "init", "WantedBy": "multi-user.target", "Wants": "network-
online.target", "WatchdogTimestamp": "Wed 2022-10-12 13:33:00 EDT", "Watch
dogTimestampMonotonic": "510245817617", "WatchdogUSec": "0"}}

PLAY RECAP **
******
rhel1.example.com   : ok=6   changed=2   unreachable=0   failed=
0   skipped=0   rescued=0   ignored=0

Figure 9.26 – Successful DHCP server playbook run

In this section, we learned how to set up a DHCP server, including static entries in the configuration files. In the next section, we will work with printers that are on your network at home or work.

Managing printers

In this section, we will be managing networked printers.

Question 1 – For this question, we would like you to set up a network printer on rhel1.example.com. Please set up a print queue named myqueue. If you have a network printer, you can test this. If not, you will not be able to test this solution/answer.

Answer 1 – We will start by installing cups on the system:

```
[emcleroy@rhel1 ~]$ sudo dnf install cups -y
```

We will then enable and start `cups` print services:

```
[emcleroy@rhel1 ~]$ sudo systemctl enable cups
[emcleroy@rhel1 ~]$ sudo systemctl start cups
```

Next, we will allow `mdns` firewall rules to allow access to the printer services:

```
[emcleroy@rhel1 ~]$ sudo firewall-cmd --permanent
--add-service=mdns
success
[emcleroy@rhel1 ~]$ sudo firewall-cmd --reload
success
```

We will use the `ippfind` command to find the printers available. In my case, I will use `Brother`:

```
[emcleroy@rhel1 ~]$ ippfind -T 5
ipp://BRW9C305BC2B044.local:631/ipp/print
ipp://BRW9C305BC2B044.local:631/ipp/print
```

Next, we will use the `ipp` address to create the print queue, `myqueue`, and add the printer to the queue:

```
[emcleroy@rhel1 ~]$ lpadmin -p myqueue -v ipp://
BRW9C305BC2B044.local:631/ipp/print -m everywhere -E
```

If you run into the `lpadmin: Unable to connect to "BRW9C305BC2B044.local:631":`
`Name or service not known` error, make sure that your DNS can look up that local address, and if not, put the host record that directs the use of a shortname or URL to an IP address in your `/etc/hosts` file.

Finally, we will set the default queue destination for printing:

```
[emcleroy@rhel1 ~]$ lpadmin -d myqueue
```

Next up, we will set up printers with Ansible Automation using the following playbook:

```
---
- name: Install cups, print queue, and printer
  hosts: rhel1.example.com
  become: true
  become_method: sudo

  tasks:
    - name: Install cups
```

```
        package:
          name: cups
          state: latest

      - name: Enable firewall rules
        firewalld:
          permanent: true
          state: enabled
          service: mdns

      - name: Reload Firewall
        command:
          cmd: firewall-cmd --reload

      - name: Start and enable cups
        service:
          name: cups
          state: started
          enabled: true

      - name: Create print queue
        command:
          cmd: lpadmin -p myqueue -v ipp://BRW9C305BC2B044.
  local:631/ipp/print -m everywhere -E

      - name: Set default print destination
        command:
          cmd: lpadmin -d myqueue
```

After we have created the playbook, we can run it using the `ansible-playbook -i inventory cups.yml -u emcleroy -k --ask-become -v` command to run an Ansible playbook successfully, as can be seen in the following screenshot:

```
 ⦿ ⦿ ⦿            jmcleroy — emcleroy@rhel3:~/playbooks — ssh emcleroy@rhel3 — 74×25
UnitFileState": "enabled", "UtmpMode": "init", "WantedBy": "multi-user.tar
get", "WatchdogTimestamp": "Wed 2022-10-12 15:37:55 EDT", "WatchdogTimesta
mpMonotonic": "517740189195", "WatchdogUSec": "0"}}

TASK [Create print queue] *********************************************
******
changed: [rhel1.example.com] => {"changed": true, "cmd": ["lpadmin", "-p",
 "myqueue", "-v", "ipp://BRW9C305BC2B044.local:631/ipp/print", "-m", "ever
ywhere", "-E"], "delta": "0:00:01.576585", "end": "2022-10-12 16:36:49.985
287", "rc": 0, "start": "2022-10-12 16:36:48.408702", "stderr": "", "stder
r_lines": [], "stdout": "", "stdout_lines": []}

TASK [Set default print destination] **********************************
******
changed: [rhel1.example.com] => {"changed": true, "cmd": ["lpadmin", "-d",
 "myqueue"], "delta": "0:00:00.008261", "end": "2022-10-12 16:36:50.356057
", "rc": 0, "start": "2022-10-12 16:36:50.347796", "stderr": "", "stderr_l
ines": [], "stdout": "", "stdout_lines": []}

PLAY RECAP ************************************************************
******
rhel1.example.com          : ok=7    changed=3    unreachable=0    failed=
0    skipped=0    rescued=0    ignored=0

[emcleroy@rhel3 playbooks]$ ▯
```

Figure 9.27 – Successful cups playbook run

In this section, we learned how to connect to a network printer that may be available to you to utilize for your everyday work. In the next section, we will show you how to set up email services on your servers.

Managing email services

In this section, we will be creating email services on a server.

Question 1 – For this question, we would like you to make a null zero client on rhel1.example. com. We would like any locally delivered mail to be forwarded to rhel2.example.com for normal delivery to the mail services:

```
myorigin: example.com
relayhost: rhel2.example.com
```

Answer 1 – We will start by installing postfix. We will then set the correct parameters that are noted in the question and ensure that the mail server does not accept external or internal mail:

```
[root@rhel1 ~]# sudo dnf install postfix -y
[root@rhel1 ~]# sudo postconf -e "inet_interfaces = loopback-only"
[root@rhel1 ~]# sudo postconf -e "myorigin = example.com"
[root@rhel1 ~]# sudo postconf -e "inet_protocols = ipv4"
[root@rhel1 ~]# sudo postconf -e "mydestination ="
[root@rhel1 ~]# sudo postconf -e "mynetworks = 127.0.0.0/8"
[root@rhel1 ~]# sudo postconf -e "local_transport = error: no local delivery"
[root@rhel1 ~]# sudo postconf -e "relayhost = [rhel2.example.com]"
[root@rhel1 ~]# sudo systemctl start postfix
[root@rhel1 ~]# sudo systemctl enable postfix
Created symlink /etc/systemd/system/multi-user.target.wants/
postfix.service → /usr/lib/systemd/system/postfix.service.
```

Keep in mind that if the exam has you set this for IPv6 as well, make sure to set up inet_protocols for both IPv4 and IPv6.

Next, we will create an Ansible playbook to complete this same process:

```
---
- name: Configure Null Client Email Service
  become: true
  hosts: rhel1.example.com
  vars:
    postfix_conf:
      relayhost: "[??.example.com]"
      inet_interfaces: "loopback-only"
      mynetworks: "127.0.0.0/8"
      myorigin: "example.com"
      mydestination: ""
  roles:
    - linux-system-roles.postfix
```

After we have created the playbook, we will use the `ansible-playbook -i inventory email.yml -u emcleroy -k --ask-become -v` command and run an Ansible playbook successfully, as shown in the following screenshot:

```
jmcleroy — emcleroy@rhel3:~/playbooks — ssh emcleroy@rhel3 — 74×25

eDirectoryMode": "0755", "RuntimeDirectoryPreserve": "no", "RuntimeMaxUSec
": "infinity", "SameProcessGroup": "no", "SecureBits": "0", "SendSIGHUP":
"no", "SendSIGKILL": "yes", "Slice": "system.slice", "StandardError": "inh
erit", "StandardInput": "null", "StandardInputData": "", "StandardOutput":
 "journal", "StartLimitAction": "none", "StartLimitBurst": "5", "StartLimi
tIntervalUSec": "10s", "StartupBlockIOWeight": "[not set]", "StartupCPUSha
res": "[not set]", "StartupCPUWeight": "[not set]", "StartupIOWeight": "[n
ot set]", "StateChangeTimestamp": "Wed 2022-10-12 16:54:00 EDT", "StateCha
ngeTimestampMonotonic": "522305625915", "StateDirectoryMode": "0755", "Sta
tusErrno": "0", "StopWhenUnneeded": "no", "SubState": "running", "SuccessA
ction": "none", "SyslogFacility": "3", "SyslogLevel": "6", "SyslogLevelPre
fix": "yes", "SyslogPriority": "30", "SystemCallErrorNumber": "0", "TTYRes
et": "no", "TTYVHangup": "no", "TTYVTDisallocate": "no", "TasksAccounting"
: "yes", "TasksCurrent": "3", "TasksMax": "11489", "TimeoutStartUSec": "1m
in 30s", "TimeoutStopUSec": "1min 30s", "TimerSlackNSec": "50000", "Transi
ent": "no", "Type": "forking", "UID": "[not set]", "UMask": "0022", "UnitF
ilePreset": "disabled", "UnitFileState": "enabled", "UtmpMode": "init", "W
antedBy": "multi-user.target", "WatchdogTimestamp": "Wed 2022-10-12 16:54:
00 EDT", "WatchdogTimestampMonotonic": "522305625914", "WatchdogUSec": "0"
}}

PLAY RECAP ******************************************************************
******
rhel1.example.com          : ok=6    changed=4    unreachable=0    failed=
0    skipped=0    rescued=0    ignored=0
```

Figure 9.28 – Successful Ansible playbook run for email services

In this section, we showed you how to modify email services and provided an example of setting up a null client configuration. In the next section, we will be going over MariaDB configuration and administration.

Managing a MariaDB database server

In this section, we will be installing MariaDB, adding to tables, backing up, restoring the content that was previously stored in the columns and tables, and searching through the database.

Question 1 – For this question, we will install and securely set up MariaDB so that root can only access locally and the root password is redhat. We will also do the following:

- Create a database named available_stock
- Create the user fred with the password redhat with full user rights on database inventory
- Create a backup of the database inventory
- Restore the backup of the database inventory

Answer 1 – We will start by installing the mariadb server, starting and enabling it, after which we will secure the installation:

```
[emcleroy@rhel1 ~]$ sudo dnf install @mariadb -y
[root@rhel1 ~]# systemctl start mariadb
[root@rhel1 ~]# systemctl enable mariadb
Created symlink /etc/systemd/system/mysql.service → /usr/lib/
systemd/system/mariadb.service.
Created symlink /etc/systemd/system/mysqld.service → /usr/lib/
systemd/system/mariadb.service.
Created symlink /etc/systemd/system/multi-user.target.wants/
mariadb.service → /usr/lib/systemd/system/mariadb.service.
[root@rhel1 ~]# mysql_secure_installation
```

We will choose to only allow root locally during the secure installation. We will choose to remove the test database. We will set a password for root at this time as well.

After we have started MariaDB and secured the installation, we will log in and create the database and user:

```
[root@rhel1 ~]# mysql -u root -p
MariaDB [(none)]> CREATE DATABASE available_stock;
Query OK, 1 row affected (0.000 sec)
MariaDB [(none)]> USE  available_stock;
Database changed
MariaDB [available_stock]> CREATE USER fred@localhost
identified by 'redhat';
Query OK, 0 rows affected (0.000 sec)
MariaDB [available_stock]> GRANT INSERT, UPDATE, DELETE, SELECT
on available_stock.* to fred@localhost;
Query OK, 0 rows affected (0.000 sec)
```

Next, we are going to take a backup of the database:

```
[root@rhel1 ~]# mysqldump available_stock -u root -p
> test.dump
Finally, to restore the dump we are going to simply change the
direction of the symbol:
[root@rhel1 ~]# mysqldump -u root -p available_stock
< test.dump
```

For the Ansible playbook, we are going to modify this, and in a slightly different way, we will set up just the database. We will now set up the MariaDB database using an Ansible Automation playbook as follows:

```
---
- name: MariaDB install and configuration
  hosts: rhel2.example.com
  become: true
  become_method: sudo

  tasks:
    - name: Install MariaDB server
      package:
        name: '@mariadb:10.3/server'
        state: present

    - name: Install MariaDB client
      package:
        name: mariadb
        state: latest

    - name: Start and enable MariaDB
      service:
        name: mariadb
        state: started
        enabled: true

    - name: Open firewall rules for MariaDB
      firewalld:
```

```
          service: mysql
          permanent: true
          state: enabled


    - name: Reload firewall
      command:
          cmd: firewall-cmd --reload


    - name: Set root password for MariaDB
      mysql_user:
          name: root
          host_all: true
          update_password: always
          password: redhat
        no_log: true
        ignore_errors: true
```

After we have completed creating the playbook, we will run the `ansible-playbook -i inventory mariadb.yml -u emcleroy -k --ask-become -v` command and see the successful output of the playbook in the following screenshot:

Figure 9.29 – Successful MariaDB playbook run

In this section, we learned how to configure MariaDB, including setting up users for the database. In the next section, we will be going over setting up web servers.

Managing web access

In this section, we will be setting up web hosts using Apache.

Question 1 – For this question, you will need to set up an Apache web host that displays **This server is working on Apache!!!** when you navigate to the rhel1.example.com server on the browser via port 80.

Answer 1 – We will start by installing Apache via the httpd package on the rhel1.example.com server using the following command:

```
[emcleroy@rhel1 ~]$ sudo dnf install httpd -y
```

After httpd has been successfully installed, we will create the index.html file in the default web page location of /var/www/html/index.html. Use your text editor of choice and create the file with **This server is working on Apache!!!** in it. Once we have created the file, we will then start and enable the service as displayed in the following commands:

```
[emcleroy@rhel1 html]$ sudo systemctl enable httpd
[emcleroy@rhel1 html]$ sudo systemctl start httpd
```

After we have started and enabled the service, we will need to open the firewall rules to allow connectivity via HTTP using the following commands:

```
[emcleroy@rhel1 html]$ sudo firewall-cmd --permanent
--add-service=http
success
[emcleroy@rhel1 html]$ sudo firewall-cmd --reload
Success
```

Once we have opened the firewall rules, we will then be able to confirm that this is working by navigating to the web page via a browser. The following screenshot demonstrates a successful deployment of Apache httpd:

Figure 9.30 – Successful web host serving the index.html file

Now, we will set up Apache `httpd` via Ansible Automation using the following playbook:

```
---
- name: Install and configure apache
  hosts: rhel1.example.com
  become: true
  become_method: sudo

  tasks:
    - name: Install apache httpd
      package:
        name: httpd
        state: latest


    - name: Ensure default directory exists
      file:
        path: /var/www/html
```

```
            state: directory
            recurse: yes

    - name: Copy index template
      template:
        src: "{{ playbook_dir }}/index.html.j2"
        dest: /var/www/html/index.html

    - name: Restore SELinux contexts
      sefcontext:
        target: /var/www/html
        setype: httpd_sys_content_t
        state: present

    - name: Open firewall rules
      command:
        cmd: firewall-cmd –permanent –add-service=http

    - name: Reload firewall
      command:
        cmd: firewall-cmd --reload

    - name: Start and enable apache httpd service
      service:
        name: httpd
        state: restarted
        enabled: true
```

Please ensure that `index.html.j2` is within your playbook directory so that it can be copied over successfully.

We will use the following command to run the playbook:

```
ansible-playbook -i inventory apache_server.yml -u emcleroy -k
--ask-become -v
```

In this section, we set up web server access. In the next section, we will be working with NFS file shares.

Managing NFS

In this section, we will be going over NFS.

Question 1 – For this question, you will need to make an NFS export on `rhel1.example.com` of the `/test` folder with read-only privileges to the `192.168.1.0/24` network and no access for the `172.16.1.0/24` network.

Question 2 – For this question, you will need to mount `/test` from `rhel1.example.com` on `rhel2.example.com` under `/mnt/test`, and you will need to ensure that it mounts on startup.

Answer 1 – We will start by installing `nfs-utils` using the following command:

```
[emcleroy@rhel1 ~]$ sudo dnf install nfs-utils
```

After we install `nfs-utils`, we will need to enable and start the service. Then, we will ensure that it is running by using the following commands:

```
[emcleroy@rhel1 ~]$ sudo systemctl enable nfs-server
Created symlink /etc/systemd/system/multi-user.target.wants/
nfs-server.service → /usr/lib/systemd/system/nfs-server.
service.
[emcleroy@rhel1 ~]$ sudo systemctl start nfs-server
[emcleroy@rhel1 ~]$ sudo systemctl status nfs-server
● nfs-server.service - NFS server and services
    Loaded: loaded (/usr/lib/systemd/system/nfs-server.service;
enabled; vendor >
    Active: active (exited) since Thu 2022-10-27 12:33:28 EDT;
10s ago
   Process: 71257 ExecStart=/bin/sh -c if systemctl -q is-active
gssproxy; then >
   Process: 71245 ExecStart=/usr/sbin/rpc.nfsd (code=exited,
status=0/SUCCESS)
   Process: 71244 ExecStartPre=/usr/sbin/exportfs -r
(code=exited, status=0/SUCC>
 Main PID: 71257 (code=exited, status=0/SUCCESS)

Oct 27 12:33:28 rhel1.example.com systemd[1]: Starting NFS
server and services.>
Oct 27 12:33:28 rhel1.example.com systemd[1]: Started NFS
server and services.
lines 1-10/10 (END)
```

After that, we will create the /test directory using the following command:

```
[emcleroy@rhel1 ~]$ sudo mkdir /test
```

After that, we will add the folder to the /etc/exports file in order to export it with the proper privileges, allow access from the 192.168.1.0/24 domain, and, by default, block it from the rest of the networks, as shown in the following screenshot:

Figure 9.31 – exports file for NFS shares

After we add the folder to the exports list, we will use the following commands to expose the NFS share:

```
[emcleroy@rhel1 /]$ sudo exportfs -rav
exporting 192.168.1.0/24:/test
```

Then, we will open the firewall rules to allow NFS services out to the world using the following commands:

```
[emcleroy@rhel1 /]$ sudo firewall-cmd --permanent --add-
service=nfs
success
[emcleroy@rhel1 /]$ sudo firewall-cmd --reload
success
```

For the Ansible solution, we will create the following playbook in order to complete the automated version of these manual tasks:

```
---
- name: Configure nfs and share folder deployment
  hosts: rhel1.example.com
  become: true
  become_method: sudo

  tasks:
    - name: Install nfs-utils
      package:
        name: nfs-utils
        state: present

    - name: Start and enable nfs-server
      service:
        name: nfs-server
        state: started
        enabled: true

    - name: Firewall rules for nfs-server
      firewalld:
        service: nfs
        permanent: true
        state: enabled

    - name: Reload firewall
      command:
        cmd: firewall-cmd --reload

    - name: Create directory
      file:
        path: /test
        state: directory
        mode: '0755'
```

```
    - name: Add directory to exports list
      lineinfile:
        path: /etc/exports
        state: present
        line: '/test 192.168.1.0/24(ro)'


    - name: Export nfs shares
      command:
        cmd: exportfs -rav
```

We will use the following command to run the `ansible-playbook -i inventory nfs_server.yml -u emcleroy -k --ask-become -v` playbook and you can see a successful run of the Ansible playbook in the following screenshot:

```
changed: [rhel1.example.com] => {"changed": true, "cmd": ["firewall-cmd", "--rel
oad"], "delta": "0:00:00.464322", "end": "2022-10-27 16:17:49.177222", "rc": 0,
"start": "2022-10-27 16:17:48.712900", "stderr": "", "stderr_lines": [], "stdout
": "success", "stdout_lines": ["success"]}

TASK [Create directory] ***************************************************
ok: [rhel1.example.com] => {"changed": false, "gid": 0, "group": "root", "mode":
 "0755", "owner": "root", "path": "/test", "secontext": "unconfined_u:object_r:d
efault_t:s0", "size": 22, "state": "directory", "uid": 0}

TASK [Add directory to exports list] **************************************
changed: [rhel1.example.com] => {"backup": "", "changed": true, "msg": "line add
ed"}

TASK [Export nfs shares] **************************************************
changed: [rhel1.example.com] => {"changed": true, "cmd": ["exportfs", "-rav"], "
delta": "0:00:00.004427", "end": "2022-10-27 16:17:50.953823", "rc": 0, "start":
 "2022-10-27 16:17:50.949396", "stderr": "", "stderr_lines": [], "stdout": "expo
rting 192.168.1.0/24:/test", "stdout_lines": ["exporting 192.168.1.0/24:/test"]}

PLAY RECAP ****************************************************************
rhel1.example.com          : ok=8    changed=3    unreachable=0    failed=0    s
kipped=0    rescued=0    ignored=0

[emcleroy@rhel3 playbooks]$ 
```

Figure 9.32 – Successful playbook run creating an NFS share

Answer 2 – After we have completed *Question 1*, we can dive into attaching the share folder to a new machine. On Rhel2, we will start by opening the firewall to allow NFS connectivity, as shown in the following commands:

```
[emcleroy@rhel2 ~]$ sudo firewall-cmd --permanent --add-
service=nfs
success
[emcleroy@rhel2 ~]$ sudo firewall-cmd --reload
success
```

Next, we are going to create the mount point folder in order to add the share to /etc/fstab and mount it:

```
[emcleroy@rhel2 ~]$ sudo mkdir /mnt/test
```

After that, we are going to add the NFS share to the /etc/fstab file in order to set it to mount at startup. The file should look something like the following screenshot but may differ depending on how you set up your test environment:

```
#
# /etc/fstab
# Created by anaconda on Thu Oct  6 00:04:47 2022
#
# Accessible filesystems, by reference, are maintained under '/dev/disk/'.
# See man pages fstab(5), findfs(8), mount(8) and/or blkid(8) for more info.
#
# After editing this file, run 'systemctl daemon-reload' to update systemd
# units generated from this file.
#
/dev/mapper/rhel-root   /                           xfs     defaults        0 0
UUID=483cae02-6952-42f8-9428-623b09aa8243 /boot                     xfs     defaul
ts          0 0
/dev/mapper/rhel-swap    swap                       swap    defaults        0 0
192.168.1.198:/test     /mnt/test   []              nfs     defaults        0 0




-- INSERT --
```

Figure 9.33 – fstab showing the mounts to run at startup

Once this has been added to fstab, we will then mount the items in fstab using the following command:

```
[emcleroy@rhel2 ~]$ sudo mount -a -v
/                           : ignored
```

```
/boot                        : already mounted
swap                         : ignored
/mnt/test                    : mounted
```

Now, to test that this is working, add a file to the `test` directory on `rhel1` using the following command:

```
[emcleroy@rhel1 /]$ sudo touch /test/test.txt
```

Once you have done that, navigate to the folder mount point of `/mnt/test` and ensure you can view the file from `rhel2` using the following commands:

```
[emcleroy@rhel2 ~]$ cd /mnt/test
[emcleroy@rhel2 test]$ ls -l
total 0
-rw-r--r--. 1 root root 0 Oct 27 13:05 test.txt
[emcleroy@rhel2 test]$
```

As you can see, you are able to view the file. Now, try to add a file to the `/mnt/test` folder on `rhel2` and you should receive an error message, such as the one shown in the following screenshot, as this is a read-only share:

Figure 9.34 – Demonstration of being unable to write to any file within the share from outside of the server

Now, you have successfully mounted the read-only share, which will persist through a reboot as fstab mounts the listed mounts at startup.

Next, we will complete these manual tasks using Ansible Automation. We will use the following playbook to accomplish the mounting of the share:

```
---
- name: Mount nfs share
  hosts: rhel2.example.com
  become: true
  become_method: sudo

  tasks:
    - name: Create mount directory
      file:
        path: /mnt/test
        state: directory
        mode: '0755'

    - name: Add fstab mount information
      lineinfile:
        path: /etc/fstab
        state: present
        line: '192.168.1.198:/test /mnt/test nfs defaults 0 0'

    - name: Remount fstab shares
      command:
        cmd: mount -a -v
```

After we have created the playbook, we will run it with the `ansible-playbook -i inventory nfs_client.yml -u emcleroy -k --ask-become -v` command. The following screenshot shows the successful execution of the playbook:

```
TASK [Add fstab mount information] *************************************
changed: [rhel2.example.com] => {"backup": "", "changed": true, "msg": "line add
ed"}

TASK [Remount fstab shares] ********************************************
[WARNING]: Consider using the mount module rather than running 'mount'.  If you
need to use command because mount is insufficient you can add 'warn: false' to
this command task or set 'command_warnings=False' in ansible.cfg to get rid of
this message.

changed: [rhel2.example.com] => {"changed": true, "cmd": ["mount", "-a", "-v"],
"delta": "0:00:00.003488", "end": "2022-11-21 02:24:10.707984", "rc": 0, "start"
: "2022-11-21 02:24:10.704496", "stderr": "", "stderr_lines": [], "stdout": "/
                      : ignored\n/boot                    : already mounted\nswa
p                     : ignored\n/mnt/test                : already mounted\n/mn
t/test                : already mounted", "stdout_lines": ["/
      : ignored", "/boot                    : already mounted", "swap
          : ignored", "/mnt/test                : already mounted", "/mnt/test
             : already mounted"]}

PLAY RECAP ************************************************************
rhel2.example.com            : ok=6    changed=3   unreachable=0    failed=0    s
kipped=0    rescued=0    ignored=0

[emcleroy@rhel3 playbooks]$ 
```

Figure 9.35 – Successful playbook run to mount NFS share

Summary

In this chapter, we presented you with mock exam questions that you may be asked in some form on the exam itself. We explored what we have learned throughout this book through this comprehensive review. New ways of completing tasks were also shown in order for you to further enhance your abilities to answer the exam questions with ease. In the next chapter, we will look over some exam tips to save you both time and stress when taking the exam. We will go over different ways to approach the exam to effectively meet the requirements and pass. I look forward to providing these tips and tricks, and hope you benefit in some way from them and my own experience of taking these exams over the years.

10

Tips and Tricks to Help with the Exam

In this chapter, we will cover different tips and tricks I have learned over the years when taking Red Hat exams. Nothing from the actual exam will be shared, but only technical shortcuts and quality-of-life items that might help make your life easier. With what you have learned in this book so far and from studying the mock exam questions, I am sure you will succeed, but with a few more words to the wise, we can make test day less stressful. These are just some things that you can do to make life easier during the exam but are not necessary to pass it. So, without further ado, let's get started!

In this chapter, we're going to cover the following topics:

- Scheduling your exam
- Test-taking tips and tricks

Technical requirements

We will be using the setup of the systems that were noted in *Chapter 1, Block Storage – Learning How to Provision Block Storage on Red Hat Enterprise Linux*. We will utilize this setup along with the setup from the additional nics for the team exercise, which can be found in *Chapter 2, Network File Storage – Expanding Your Knowledge of How to Share Data*. This will allow us to solve all of the upcoming questions as if we were in an actual test environment.

Scheduling your exam

First things first, before we get into tips on taking the exam, you need to book it. This is done through the Red Hat exam site, which can be found at `https://www.redhat.com/en/services/certification/individual-exams`. This allows you to book the exam by either signing up and paying with two training credits or $400. You can book at a testing facility near you or you can book a remote exam. A remote exam allows you to use a laptop or desktop in a secure location. You need to keep in mind that you cannot be interrupted during the exam or they will stop it and you will forfeit the fee. A remote exam requires the following items:

💻 **System requirements**

Computer
You will need a computer with a single active monitor. Red Hat supports many Intel compatible X86_64-bit architecture computers.

USB
One USB Drive (2.0 or higher) with at least 8 GB capacity.
Note: The entire disk will be overwritten, so make sure you have saved any contents you may have on the disk before following the procedure for creating the live exam environment.

USB hub
One wired USB hub is allowed if a hub is needed to accommodate permitted peripheral devices as described below.

Hard drive
A hard drive with free storage capacity of at least 4 GB (for Live USB creation only).

Mouse
A wired mouse is optional but recommended. A wireless mouse is not allowed. A wired mouse is required if you use a laptop in a closed and docked mode as described below.

Keyboard
Only one keyboard is allowed for the exam. If you wish to use an external, wired keyboard with your laptop, you will have to use the laptop closed and docked. This will require the use of an external monitor and wired mouse as well. Wireless keyboards are not permitted.

Webcam
One external webcam with at least a 1m cable _and_minimum_720p_resolution_.

Monitors
Only one physical display will be allowed for the exam.

Connecting an external monitor to a laptop
You are only allowed to use one monitor, one keyboard and one external mouse. If you chose to connect an external monitor to your laptop, the laptop lid must be closed throughout the duration of the exam session. You will be required to use a wired keyboard and a wired mouse.

Sound and microphone
A working microphone is required. Verify that the audio and microphone are not set on mute prior to the exam.

Operating system
N/A.

Firewalls
Firewalls that allow normal web activities will typically work. More restrictive firewalls that limit outgoing access and that require additional authentication may cause problems. Most firewall issues will show up when you run the compatibility test.

RAM
Minimum 4GB of RAM are required.

Internet connection speed
Download speed requirements are 768Kbps and upload speed requirements are 512Kbps.

Network connection
Unless it is physically not possible, a wired network connection should be used, not wireless, to ensure the most reliable delivery of your exam.

Figure 10.1 – Remote exam requirements

In the following screenshot, you can see the remaining requirements for taking a remote exam:

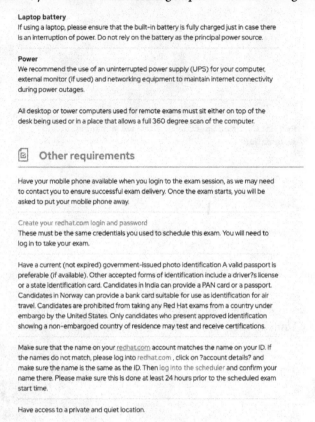

Laptop battery
If using a laptop, please ensure that the built-in battery is fully charged just in case there is an interruption of power. Do not rely on the battery as the principal power source.

Power
We recommend the use of an uninterrupted power supply (UPS) for your computer, external monitor (if used) and networking equipment to maintain internet connectivity during power outages.

All desktop or tower computers used for remote exams must sit either on top of the desk being used or in a place that allows a full 360 degree scan of the computer.

🗓 **Other requirements**

Have your mobile phone available when you login to the exam session, as we may need to contact you to ensure successful exam delivery. Once the exam starts, you will be asked to put your mobile phone away.

Create your redhat.com login and password
These must be the same credentials you used to schedule this exam. You will need to log in to take your exam.

Have a current (not expired) government-issued photo identification A valid passport is preferable (if available). Other accepted forms of identification include a driver?s license or a state identification card. Candidates in India can provide a PAN card or a passport. Candidates in Norway can provide a bank card suitable for use as identification for air travel. Candidates are prohibited from taking any Red Hat exams from a country under embargo by the United States. Only candidates who present approved identification showing a non-embargoed country of residence may test and receive certifications.

Make sure that the name on your redhat.com account matches the name on your ID. If the names do not match, please log into redhat.com , click on ?account details? and make sure the name is the same as the ID. Then log into the scheduler and confirm your name there. Please make sure this is done at least 24 hours prior to the scheduled exam start time.

Have access to a private and quiet location.

Figure 10.2 – Additional remote exam requirements

After you have purchased the exam, you can use the scheduler to set up your exam either remotely or at a facility near you. As you can see here in the scheduler, I have one exam that is available to schedule:

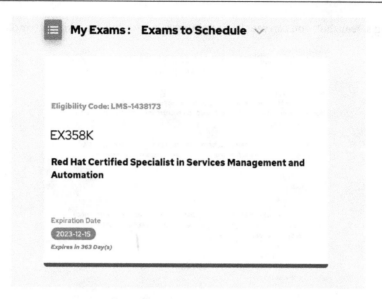

Figure 10.3 – Exam available to schedule in the Red Hat exam scheduler

After I choose the exam to schedule, I am then presented with the following screen with options for a location, such as the United States, and what type of exam I would like to take, either at a facility or remotely:

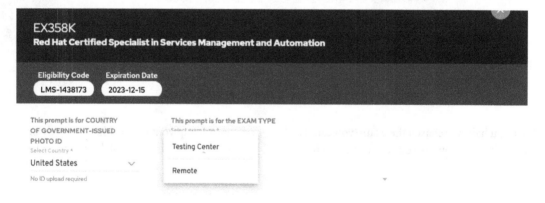

Figure 10.4 – Choosing your exam type

In my case, I have chosen the remote option for taking the exam:

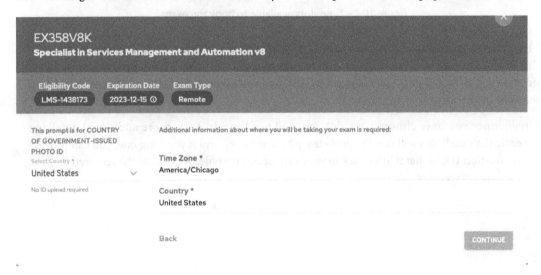

Figure 10.5 – Remote exam setup

The following screenshot showcases the next items you need to provide for setting up the remote exam:

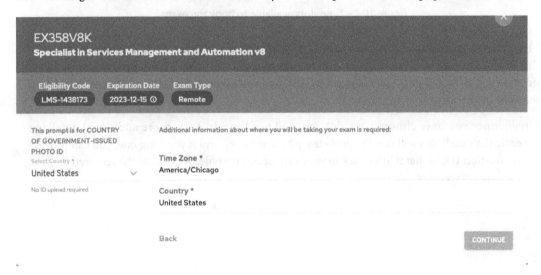

Figure 10.6 – Additional information for scheduling the exam

After you have provided the required information, you will then be taken to the exam scheduler, which shows you the available slots. Do not worry if you do not see a slot you want, if you reload the page by going through the preceding steps, there is a chance someone will have canceled and opened up a slot where you intend to take the exam. Please note that you have to cancel or reschedule your exam before you are within the 24-hour window, otherwise, you will be unable to change the time without losing your payment. In this screenshot, you can see an example of what time slots are available:

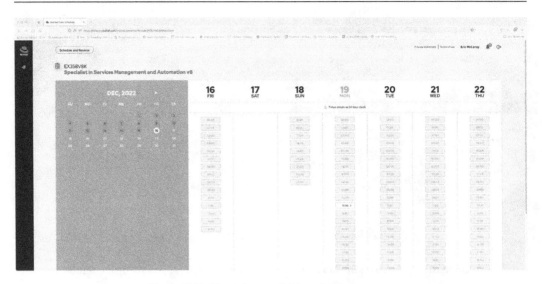

Figure 10.7 – Time slots available to book the exam

I have chosen an available slot, in my case 11:15 AM. This is a 24-hour clock, so please be aware of that when setting up your exam. If you want to start your exam at 4:00 PM, you need to choose 16:00, not 4:00, as that would put your exam at 4:00 AM, and you might be in for an early phone call if you schedule incorrectly.

Finally, once you have chosen your slot, you will need to add a few more additional details to schedule the exam. You will need to provide a phone number, and if you have one, you can provide a **Certification ID**, so that it links back to your certification profile. We can see the required fields in the following screenshot:

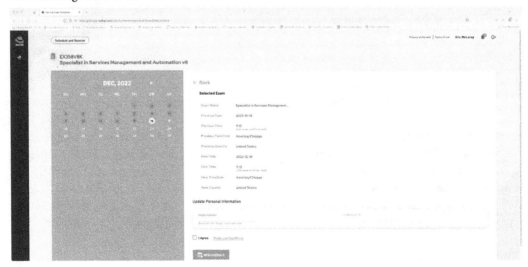

Figure 10.8 – Scheduling the exam and agreeing to the terms and conditions

Once you have gone through all of these settings and agreed to the terms and conditions, your exam will be scheduled.

On exam day, you will need to be in a location that is free of writing with no paper or pencils around and other electronic devices than your workstation. The proctor will tell you what you can and cannot have present during the exam. You will need to place your phone away from you, and on silent, in case the exam proctor needs to contact you and cannot through the onscreen prompts. You will also have to place your ID after it is verified out of sight. You can have a beverage for the duration of the exam; however, it will be inspected by the proctor, so ensure that there is no writing on the container for the best test-taking experience. You will also be allowed to use the bathroom at an interval that the proctor will provide you. Please note that if you time your bathroom break at the 1:55 hour mark, there is a chance you will not have to perform a second room check outside of the check that occurs due to taking a break.

I have set up for myself a quiet little place in one of my sheds that allows privacy and full connectivity. As recommended, I have run an Ethernet cable to the shed along with a **Uninterruptable Power Supply (UPS)** for my laptop. I find that using the laptop screen is somewhat difficult as it is smaller than I am used to, so I recommend using a monitor you are comfortable with as you will want to maximize your space as you are only allowed one monitor. This means if you are using a laptop, it will have to remain closed while connected to the monitor, as you are not permitted to use two monitors.

Once you have set up a quiet place where you can safely take the exam, you should run the compatibility test within the exam environment (this image link will be emailed to you once you have scheduled your exam) to ensure that all of your equipment meets the recommended setup. If you opt to use the laptop screen with a built-in camera, ensure that it is enabled. It is much easier for the proctor to see everything and for you to be able to position the wired camera more easily if you have both of these webcams enabled. The compatibility information can be found in the following screenshot:

Run a compatibility test in the remote exam live environment ISO

In advance of your exam, you are required to run a compatibility check in the remote exam live environment ISO on the hardware and network that you are planning to use for the exam, preferably at the same time of day as the scheduled exam. You can only proceed with your scheduled Red Hat remote exam if your compatibility test is successful.

Once you have booted into the remote exam live environment ISO, you will be prompted with an option to run a compatibility check or access upcoming exams.

If the compatibility test is successful, good luck on your exam!

If the compatibility test is unsuccessful and your exam date and time is more than 24 hours away, please cancel your exam and reschedule after the issues are resolved.

If the compatibility test is unsuccessful and your exam is in the next 24 hours:

* Do not cancel your exam, as you will not be able to reschedule your exam
* Contact our team on chat and describe the issues you are experiencing

Once you have booted into the remote exam live environment ISO, you will be prompted with an option to run a compatibility check or access upcoming exams.

Please note that office, home, and other networks can vary widely in their available bandwidth depending on the time of day, so please make sure to run the compatibility test at the time of day your exam is scheduled to get the most accurate evaluation possible. We recommend running the compatibility test two to three days prior to your scheduled exam date. Please note that we have configured and restricted the remote exam live environment for the sole purpose of taking the Red Hat exam and running the compatibility test.

Important compatibility information

Warnings of incompatibility: In the event that the compatibility tool warns about a condition that might pose a problem, please heed this guidance and find ways to address the issue. For example, if networking does not appear to be capable of providing a reliable delivery, please find another location and network and run the compatibility test again.

Please note that a successful run of the compatibility tool does not guarantee the absence of issues when your exam is delivered. Because you will use your device, location and network, we cannot troubleshoot these for you. In the event there are problems that significantly impede your ability to take the exam, we will typically discontinue the exam and allow you to reschedule.

You will not be able to take your exam remotely unless you meet all of the above requirements.

Otherwise, you will need to reschedule your exam or take the exam at a nearby testing center. Please refer to the section below, ?how to cancel or reschedule a scheduled exam? for additional information.

Figure 10.9 – Compatibility test information for a remote exam

After you have run the compatibility test and ensured that everything meets the requirements, you can then relax and know that you are safe to take the exam at the scheduled time and date. This will allow you to mentally prepare for the exam by not having to worry about your setup during the exam that you are taking remotely.

On test day, you will be able to log in using your Red Hat login and start your scheduled exam at the time you have selected. The proctor will then instruct you on what you need to do during the exam. This will allow you to understand everything that is required of you during the remote exam, thus enabling your success at taking the exam at your convenience within your own home.

Once that is all squared away, we will follow that setup with some tips on taking the exam itself, which will help you feel more at ease and comfortable.

Tips and tricks that can help while taking the test

First, let's start with something that can make your life easier right from the start. Read all of the directions from top to bottom. The reason I say this is because it can save you time if it tells you something specific that can be omitted from repetition, such as only doing things on certain servers. It will also let you know how things will be graded. By reading the directions, you will understand your test environment better and be prepared to dive head-first into the material. Anywhere you can save time is a win.

Speaking of time, utilize the full 4 hours for the exam. Do not try and rush it to wrap it up sooner; you are not graded on the time it takes other than that 4-hour window, which is all the time you are allotted for the exam. It will shut down your access to the exam when the timer hits 0. Do not let a question rob you of time. If you are better at other items, do those first and come back to the tougher questions if you have the time so that you can cover the most ground.

Another thing to keep in mind for saving time is to do everything that is not specifically called out to a specific user as root. You can always restore the SELinux contexts with a `restorecon -Rv <directory>` that may receive the wrong context due to the user. By using root, you will not run into issues where you might forget to use the sudo command, and being the root user directly makes finding files a breeze as you can access the entire file structure. Just be careful, as this can lead to permission issues with files and folders, so if the exam tells you to use a specific user for an action, do so. This is not best practice in everyday life but can be a time saver on the exam.

Next, if you are used to using the GUI interface, for instance, and the CLI isn't as quick for you, there is something you can do to make your life easier. You can install the GUI on the servers you will be making changes to. Let's go through how to install the GUI on a RHEL 8.1 server so that you have it up your sleeve if it is better suited for you.

First, we are going to install the *Server with GUI* package on the servers we would like to have the GUI on. They can be run concurrently, so you do not have to wait on one server to finish to do the next, as shown in the following screenshots:

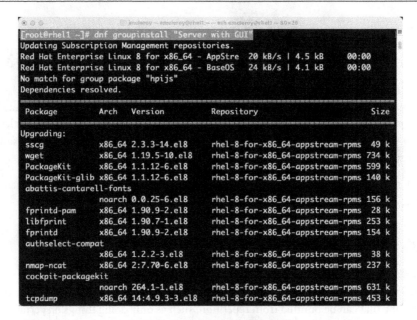

Figure 10.10 – Installation of the Server with GUI commands

After you run the command to install the *Server with GUI* package, you will see the end result, as shown in the following screenshot:

```
nfs-utils-1:2.3.3-51.el8.x86_64
bluez-libs-5.56-3.el8.x86_64
python3-cloud-what-1.28.29-3.el8.x86_64
libqmi-1.30.2-1.el8.x86_64
libqmi-utils-1.30.2-1.el8.x86_64
libmbim-1.26.0-2.el8.x86_64
xmlrpc-c-client-1.51.0-6.el8.x86_64
fuse-common-3.3.0-15.el8.x86_64
ndctl-libs-71.1-3.el8.x86_64
bluez-obexd-5.56-3.el8.x86_64
rhsm-icons-1.28.29-3.el8.noarch
xmlrpc-c-1.51.0-6.el8.x86_64
ndctl-71.1-3.el8.x86_64
python3-nftables-1:0.9.3-25.el8.x86_64
glibc-gconv-extra-2.28-189.5.el8_6.x86_64
NetworkManager-initscripts-updown-1:1.36.0-7.el8_6.noarch
kernel-modules-4.18.0-372.26.1.el8_6.x86_64
libatomic-8.5.0-10.1.el8_6.x86_64
mtools-4.0.18-15.el8_6.x86_64
systemd-container-239-58.el8_6.7.x86_64
iproute-tc-5.15.0-4.el8_6.1.x86_64
kernel-core-4.18.0-372.26.1.el8_6.x86_64

Complete!
[root@rhel1 ~]#
```

Figure 10.11 – Completed install of the GUI

We are then going to set it so that the server boots from the graphical interface and reboots so you can access it from the console, as shown in the following screenshot:

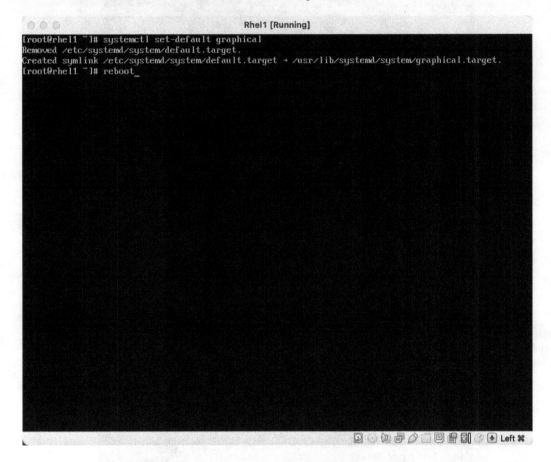

Figure 10.12 – Commands to set the graphical interface as the default startup

After the reboot, it should load up into the GUI, as shown in the following screenshot:

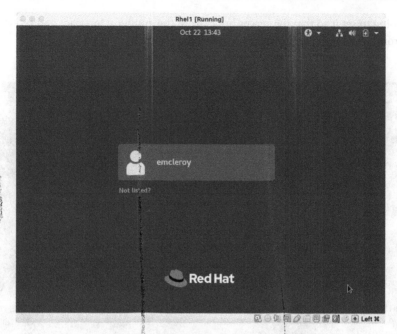

Figure 10.13 – GUI login screen for RHEL 8.1

The next thing I want you to remember is that if you get stumped, you can not only access the **manual pages (man)** pages, but you can also access documentation found in the /usr/share folder, as shown in the following screenshot:

Figure 10.14 – Folder that contains the installed package documentation

This directory holds information on packages that have been installed on the system. So, if you are stuck on something such as the **Domain Name System** (**DNS**) or the **Dynamic Host Configuration Protocol** (**DHCP**), you can look up information on the subject to get your mind rolling again. This can be extremely helpful when you feel you have forgotten everything you learned before sitting down to take the exam.

Now, onto man pages, which contain more detailed information on the commands and processes you can perform with the different packages you have installed. You can also do a search for items found in the man pages directory, and when you see a gzip in the `/usr/share/man/` directory, this will provide you with the name of the man page you need to access for further information. We can see a search for DHCP information in the following screenshot:

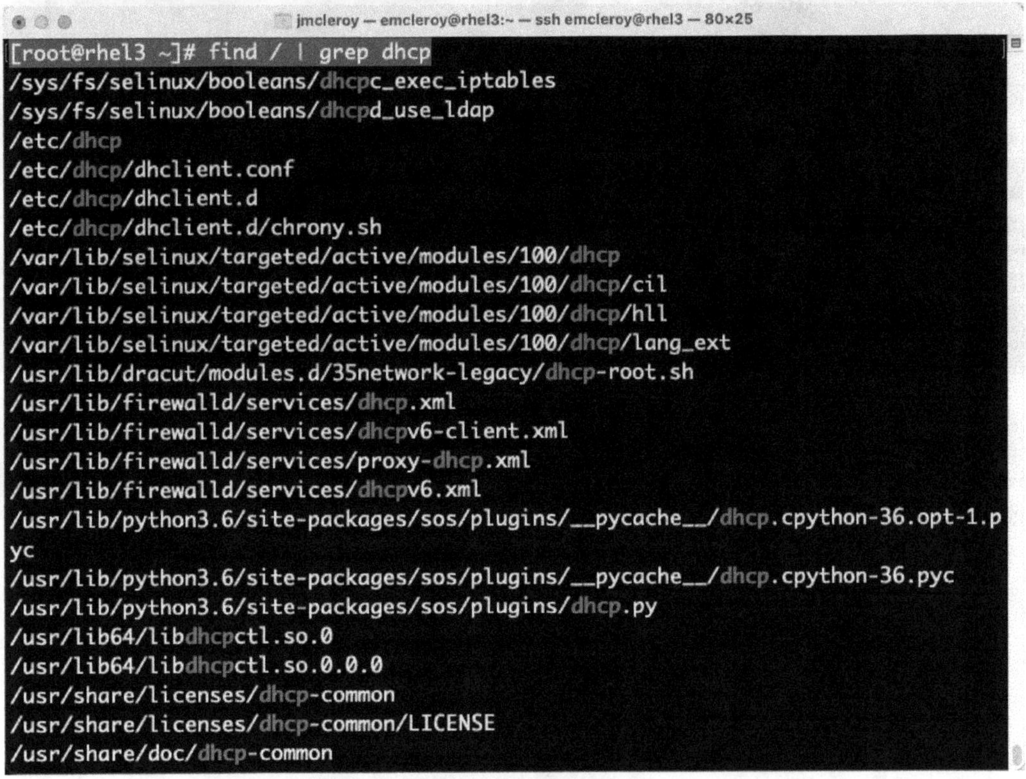
jmcleroy — emcleroy@rhel3:~ — ssh emcleroy@rhel3 — 80×25
[root@rhel3 ~]# find / | grep dhcp
/sys/fs/selinux/booleans/dhcpc_exec_iptables
/sys/fs/selinux/booleans/dhcpd_use_ldap
/etc/dhcp
/etc/dhcp/dhclient.conf
/etc/dhcp/dhclient.d
/etc/dhcp/dhclient.d/chrony.sh
/var/lib/selinux/targeted/active/modules/100/dhcp
/var/lib/selinux/targeted/active/modules/100/dhcp/cil
/var/lib/selinux/targeted/active/modules/100/dhcp/hll
/var/lib/selinux/targeted/active/modules/100/dhcp/lang_ext
/usr/lib/dracut/modules.d/35network-legacy/dhcp-root.sh
/usr/lib/firewalld/services/dhcp.xml
/usr/lib/firewalld/services/dhcpv6-client.xml
/usr/lib/firewalld/services/proxy-dhcp.xml
/usr/lib/firewalld/services/dhcpv6.xml
/usr/lib/python3.6/site-packages/sos/plugins/__pycache__/dhcp.cpython-36.opt-1.p
yc
/usr/lib/python3.6/site-packages/sos/plugins/__pycache__/dhcp.cpython-36.pyc
/usr/lib/python3.6/site-packages/sos/plugins/dhcp.py
/usr/lib64/libdhcpctl.so.0
/usr/lib64/libdhcpctl.so.0.0.0
/usr/share/licenses/dhcp-common
/usr/share/licenses/dhcp-common/LICENSE
/usr/share/doc/dhcp-common

Figure 10.15 – Searching for files with DHCP in them

You can see in the folder in the following screenshot that you have multiple man pages that contain DHCP information:

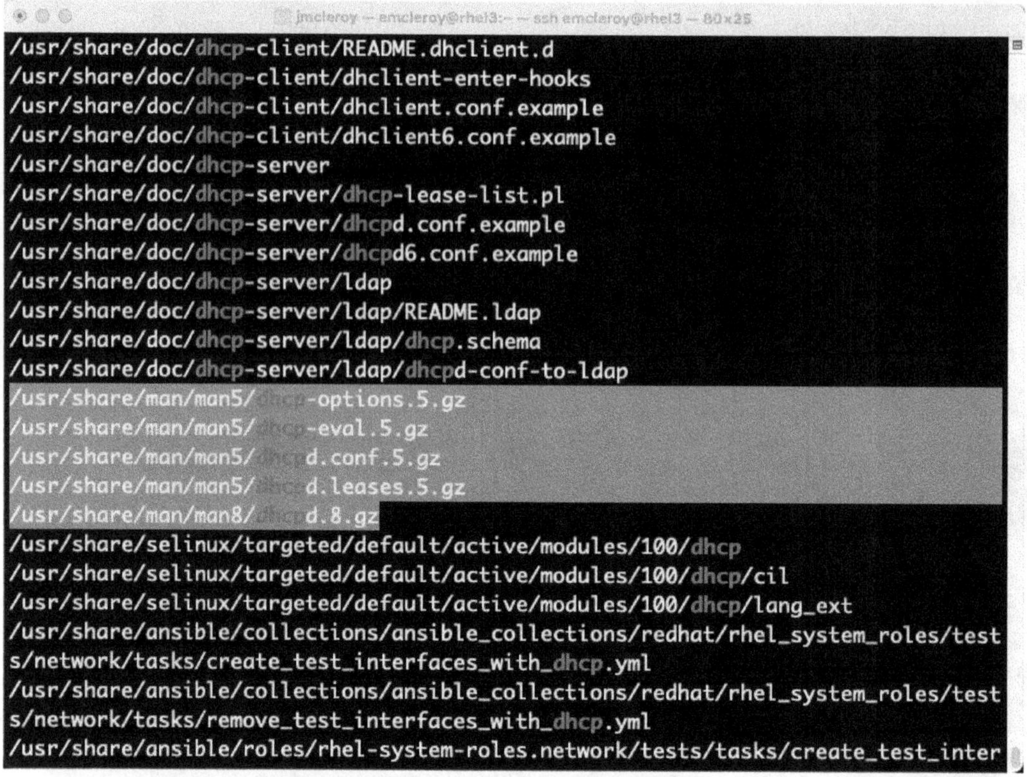

Figure 10.16 – DHCP man page names

Now, you can see that there are multiple man pages with dhcpd in them, and you can open a man page for further documentation, as shown in the following screenshot:

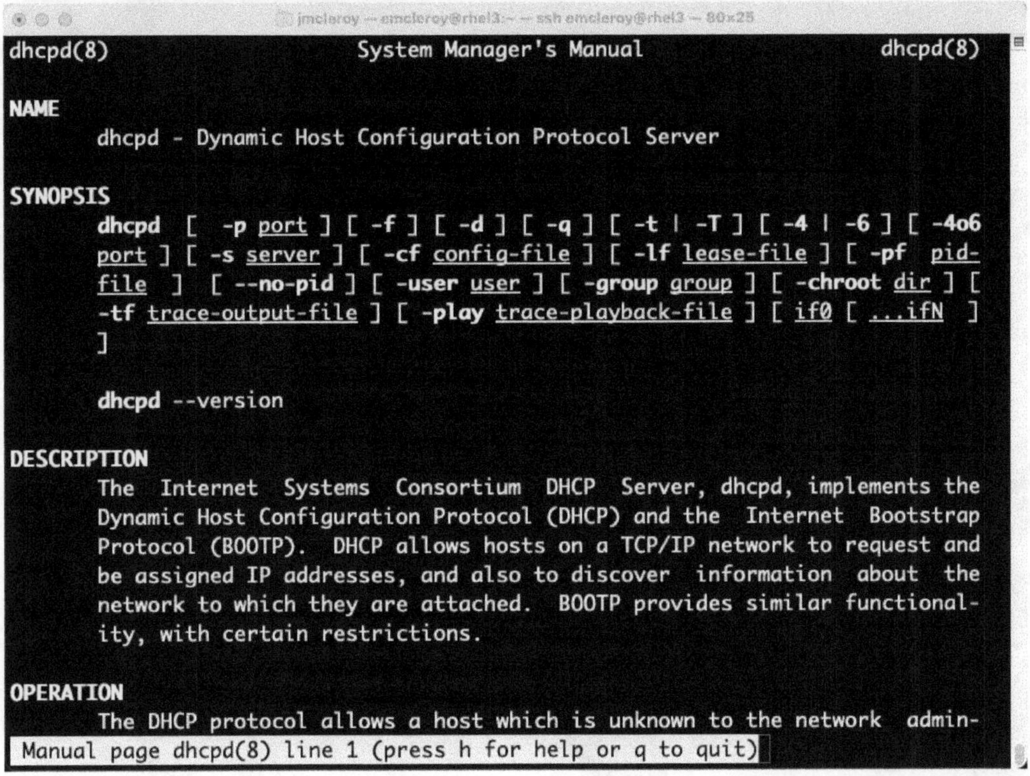

Figure 10.17 – A dhcpd man page

These gzip files help if you are stumped on the name of the man pages and need a quick way to find all of the man pages with `dhcp` in them somewhere.

Please note that all of your changes must be persistent when completing the exam. They will reboot all systems and wipe the Ansible nodes so that they can be utilized for running fresh playbooks.

Summary

Finally, we are at the end of our journey through this book. I want to say thank you for choosing to go on this journey with me into the world of Red Hat Enterprise Linux. It was a pleasure to write this book with the hope that I can help as many of you as possible to succeed in your goals of becoming a certified **Red Hat Certified Specialist in Services Management and Automation** (**EX358**). This exam counts toward the five additional exams needed to receive the **Red Hat Certified Architect** (**RHCA**) along with the **Red Hat Certified Expert** (**RHCE**) exam. I hope you gained some insights into how I accomplish tasks. However, it's important you know that these are not the only ways you can complete these tasks to accomplish your certification goals, but some of the many that will hopefully spur you on to the path of victory.

Index

A

Ansible
used, for automating network services 70-78
Ansible Automation
email services, setting up via 182-186
printer services, setting up via 169-175
used, for automating traffic control 240-248
used, for automating web servers 240-248
used, for installing and configuring
  MariaDB on RHEL 8 209-217
used, for setting up link aggregation 97-104
Apache
web hosts, setting up with 295-297
Apache httpd
configuring 223-227
installing 220-222
manual page (man) pages 222
setting up, with TLS certificates 230
virtual hosts, troubleshooting 227-230

C

Code in Action (CiA) 48

D

DHCP server
setting up 283-286
DHCP server configuration
automating, to provide DHCP
  services 128-132
setting up, manually to provide
  DHCP services 119-128
Disaster Recovery (DR) 154
DNS server
setting up 274-282
DNS server configuration
automating, to provide DNS
  services 147-154
setting up, manually to provide
  DNS services 133-146
Domain Name System (DNS) 133, 319
**Dynamic Host Configuration
  Protocol (DHCP) 319**
using, at initial interface
  connectivity 113-119

E

email services 176
  creating, on server 289-291
  manual setup 178-182
  setting up, via Ansible Automation 182-186
Ethernet (ETH) 58
EX358 exam
  preparation 250

F

firewall services rules
  setting up 260-264
fully qualified collection name (FQCN) 27

G

GitHub access
  setting up 4, 34, 106, 188, 220

H

HAProxy
  installing 237-239
High Availability (HA) 79

I

Internet Printing Protocol (IPP) 160
Internet Protocol (IP) 58
**Internet Small Computer Systems
    Interface (iSCSI) 3**
iSCSI block storage
  Ansible automation playbook,
      creating 27-31
  Ansible automation playbook, using 27-31
  deployment 19-27

  knowledge test 17, 18
  manual provisioning 19-27
  overview 16

L

lab environment
  setting up 4-16
  setting up, for NFS 34, 35
  setting up, for SMB 34, 35
link aggregation 83, 84
  setting up, with Ansible Automation 97-104
**Link Aggregation Control
    Protocol (LACP) 79**
link aggregation profiles
  creating 84-97
Linux networking 58
  exploring 106
logical unit numbers (LUNs) 16

M

manual pages (man) pages 61, 318
MariaDB 188
  for data collection storage 188
  installing and configuring manually,
      on RHEL 8 188-209
  installation and configuration on RHEL
      8, with Ansible Automation 209-217
MariaDB database server
  managing 291-295
media access control (MAC) address 58

N

networked printers
  managing 286-289

Network File Storage (NFS) 33
  managing 298-304
  lab environment, setting up 34, 35
network interface controllers (NICs) 16
NetworkManager command line
      interface (nmcli) 61
NetworkManager text user
      interface (nmtui) 64
network profile
  creating 59-69
network services
  automating, with Ansible 70-78
  managing 251-260
network teaming
  setting up 270-274
network teaming configuration
  reference link 83
NFS Ansible Automation playbook
  creating 41-43
  usage 41-43
NFS network storage 35-40
NGINX web server
  installing 231-237

O

out-of-band management (OBM) 64
out-of-the-box DHCP configuration
  using 113

P

Pointer Record (PTR) 139, 282
printer services 158
  manual setup 159-169
  setting up, via Ansible Automation 169-175

R

Red Hat Enterprise Linux (RHEL) 3, 70
Red Hat exam
  scheduling 307-315
  tips and tricks 315-321
relational database management
      system (RDBMS) 188
RHEL 8
  Ansible Automation, used for installing
      and configuring on 209-217
  MariaDB, installing and configuring
      manually on 188-209

S

SELinux
  managing 265-267
Server Message Block (SMB) 33
  lab environment, setting up 34, 35
  network storage 35-40
SMB storage
  deployment 43- 49
  manual provisioning 43-49
SMB storage Ansible Automation playbook
  creating 49-54
  usage 49-54
Start of Authority (SOA) 141
static IP addresses
  setting up 106-112
storage area network (SAN) 16
Structured Query Language (SQL) 188
system processes
  managing 267-269

T

Time to Live (TTL) 141
traffic control 220
    automating, with Ansible
        Automation 240-248
    setting up, with HAProxy 237-239

V

Virtual Private Network (VPN) 120
    managing 265

W

web hosts
    setting up, with Apache 295-297
web servers
    automating, with Ansible
        Automation 240- 248
    manual setup 220, 222-237

Y

Yet Another Markup Language (YAML) 28

Packtpub.com

Subscribe to our online digital library for full access to over 7,000 books and videos, as well as industry leading tools to help you plan your personal development and advance your career. For more information, please visit our website.

Why subscribe?

- Spend less time learning and more time coding with practical eBooks and Videos from over 4,000 industry professionals

- Improve your learning with Skill Plans built especially for you

- Get a free eBook or video every month

- Fully searchable for easy access to vital information

- Copy and paste, print, and bookmark content

Did you know that Packt offers eBook versions of every book published, with PDF and ePub files available? You can upgrade to the eBook version at packtpub.com and as a print book customer, you are entitled to a discount on the eBook copy. Get in touch with us at customercare@packtpub.com for more details.

At www.packtpub.com, you can also read a collection of free technical articles, sign up for a range of free newsletters, and receive exclusive discounts and offers on Packt books and eBooks.

Other Books You May Enjoy

If you enjoyed this book, you may be interested in these other books by Packt:

Digital Forensics and Incident Response - Third Edition

Gerard Johansen

ISBN: 978-1-80323-867-8

- Create and deploy an incident response capability within your own organization
- Perform proper evidence acquisition and handling
- Analyze the evidence collected and determine the root cause of a security incident
- Integrate digital forensic techniques and procedures into the overall incident response process
- Understand different techniques for threat hunting
- Write incident reports that document the key findings of your analysis
- Apply incident response practices to ransomware attacks
- Leverage cyber threat intelligence to augment digital forensics findings

Network Automation with Go

Nicolas Leiva, Michael Kashin

ISBN: 978-1-80056-092-5

- Understand Go programming language basics via network-related examples
- Find out what features make Go a powerful alternative for network automation
- Explore network automation goals, benefits, and common use cases
- Discover how to interact with network devices using a variety of technologies
- Integrate Go programs into an automation framework
- Take advantage of the OpenConfig ecosystem with Go
- Build distributed and scalable systems for network observability

Packt is searching for authors like you

If you're interested in becoming an author for Packt, please visit authors.packtpub.com and apply today. We have worked with thousands of developers and tech professionals, just like you, to help them share their insight with the global tech community. You can make a general application, apply for a specific hot topic that we are recruiting an author for, or submit your own idea.

Share Your Thoughts

Now you've finished *Red Hat Certified Specialist in Services Management and Automation EX358 Exam Guide*, we'd love to hear your thoughts! Scan the QR code below to go straight to the Amazon review page for this book and share your feedback or leave a review on the site that you purchased it from.

https://packt.link/r/1803235497

Your review is important to us and the tech community and will help us make sure we're delivering excellent quality content.

Download a free PDF copy of this book

Thanks for purchasing this book!

Do you like to read on the go but are unable to carry your print books everywhere?

Is your eBook purchase not compatible with the device of your choice?

Don't worry, now with every Packt book you get a DRM-free PDF version of that book at no cost.

Read anywhere, any place, on any device. Search, copy, and paste code from your favorite technical books directly into your application.

The perks don't stop there, you can get exclusive access to discounts, newsletters, and great free content in your inbox daily

Follow these simple steps to get the benefits:

1. Scan the QR code or visit the link below

https://packt.link/free-ebook/9781803235493

2. Submit your proof of purchase
3. That's it! We'll send your free PDF and other benefits to your email directly

www.ingramcontent.com/pod-product-compliance
Lightning Source LLC
Chambersburg PA
CBHW062056050326
40690CB00016B/3106

9 781803 235493